W9-BZY-176

ROOFING

DESIGN CRITERIA, OPTIONS, SELECTION

R. D. Herbert III

ROOFING

DESIGN CRITERIA, OPTIONS, SELECTION

R. D. Herbert III
Illustrated by Carl W. Linde

Copyright 1989

R.S. MEANS COMPANY, INC.
CONSTRUCTION PUBLISHERS & CONSULTANTS

100 Construction Plaza
P.O. Box 800
Kingston, MA 02364-0800
(617) 585-7880

Southam
Construction
Information
Network

This book was edited by Mary Greene and David Zuniga. Typesetting was supervised by Helen Marcella. The book and jacket were designed by Norman Forgit. Illustrations by Carl Linde.

Printed in the United States of America

10   9   8   7   6

Library of Congress Catalog Number 89-152465

ISBN 0-87629-104-3

To my wife, Betty, whose patience knows no bounds.

TABLE OF
CONTENTS

APPENDICES

FOREWORD

Over 50 percent of post-construction problems can be attributed to roofing or related systems. Yet only one to two percent of the budget for an average commercial building project is typically devoted to this crucial component. As the building's protection against the elements, the roof is not only the most vulnerable part of the structure, but its integrity is essential, as it ensures the contents of the building against weather damage.

The goal of this book is to inform all who are involved in the selection, installation, maintenance, or replacement of roofing of the design criteria, options, and selection procedures required for a proper installation. The key is current and complete information about available materials and methods.

The first chapter of this book addresses the challenges of today's roofing industry, including new technology and its effect on design, project coordination, scheduling, and estimating. This chapter also outlines the basic components of a roofing system and the major types of roofing.

Chapter 2 presents the fundamental principles of sound roof construction, and the essential quality control that must take place before, during, and after the installation. Chapter 3 describes the basic types of structural roof decks and the methods for attaching the deck to the structure. This is the first of ten chapters covering the principle components of the roofing system, ranging from drainage and insulation to various roofing materials and accessories.

Chapters 14 through 16 provide guidelines for roof maintenance and restoration, warranties, insurance, and codes. At this point, the basic roofing systems, components, and associated considerations have been established. Chapter 17 follows with an effective approach to roof system selection. Finally, Chapter 18 presents the four levels of estimates, how they are performed, and the information required for each.

The appendices include a list of roofing industry organizations and a roofing questionnaire to aid in the selection process. Also included are charts and tables with installation recommendations and measurement guidelines. An extensive glossary follows.

Owners, architects, facilities managers, and roofing contractors should find this book helpful in selecting systems for new or replacement roofing. Inspectors and appraisers may gain better insight into the components and conditions that comprise a sound roofing system. The tasks of each of these parties are facilitated by proper and thorough knowledge of the many facets of roofing.

Acknowledgments

The author wishes to express appreciation to the entire staff of the National Roofing Contractors Association, who continue to be a reliable source of industry standards. Special thanks are also given to Margie Hagewood, my secretary, for her assistance in preparing the manuscript. Particular appreciation is due to David Zuniga, for his contributions in researching and assembling material for this book.

Gerald Sullivan, PE, from Springfield, Tennessee prepared Chapter Three and made many valuable suggestions to the remainder of the book.

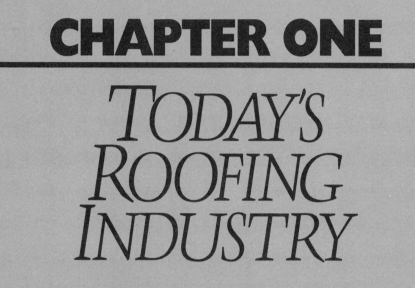

CHAPTER ONE

TODAY'S ROOFING INDUSTRY

CHAPTER ONE

Today's Roofing Industry

A Rapidly Changing Technology

"The roof is leaking!" Those are distressing words to an architect, contractor, building manager, or roofing installer —especially when voiced by an important client or describing a large or high profile project. When the roof system fails, one of these parties is inevitably responsible. This book is dedicated to informing and instructing those involved in the selection, installation, maintenance, and/or replacement of roofing to avoid the untimely failures so common in the industry.

Owners, architects, facilities managers, and installers will find this book useful in selecting roof systems for new or re-roof projects. Inspectors and appraisers will gain better insight into the components and conditions that comprise a sound roofing system.

The roofing industry has experienced an unprecedented increase in failures during the past ten years. Approximately 65 percent of all lawsuits brought against architects during one recent period originated with roofing problems. Many of these problems may be attributed to the introduction of a wide variety of radically new roofing products and systems. In single-ply roofing alone, there are over 75 manufacturers offering a total of several hundred products. Each manufacturer uses its own proprietary test methods, fasteners, adhesives, and detailing. This array of variables, and the accompanying mountains of information and specifications can overwhelm the designer, facilities manager, or roofing contractor. It is difficult for the agencies traditionally responsible for testing and setting standards [such as the American Society for Testing and Materials (ASTM), the National Bureau of Standards (NBS), Factory Mutual (FM), and others (see Appendix A for more information on agencies)] to test all of the new entrants into this rapidly changing market.

Old Versus New

The influx of materials, together with unclear materials standards, has left the designer or installer with one of two courses of action — either conservatively specify "whatever worked last time," or risk trying the latest roofing technology. In any case, it has been difficult to compare materials or system specifications. During the past few years, there have been tremendous advances in education and standards for the industry by such entities as ASTM,

Underwriters' Laboratory (UL), the National Roofing Contractors Association (NRCA), the Roofing Industry Educational Institute (RIEI), and others (see Appendix A).

Design: A Systems Approach

The roofing system does not exist in a vacuum; in order to select a roofing system, elements of the design, such as roof deck material, wall construction, and roof size, must be determined. Chapters 2 and 17 describe some of the important considerations in the roof system selection process and provide guidelines to consider before and during the design process.

Project Coordination

In the past, a lack of coordination between designers, installers, and maintenance personnel has led to field-designed or makeshift details. These details, in turn, have often led to long-term problems for owners, and lawsuits for designers. Chapter 2 addresses this issue, and introduces guidelines on project coordination, documentation, and record keeping. Chapters 15 and 16 outline measures to ensure the longevity of the roofing system.

Estimating

Because of the many different roofing methods and materials available today, estimating — both the materials and labor —for a roofing job is a growing challenge. Chapter 18 provides estimating guidelines, showing both a handwritten and a computer-generated estimate. If the formats shown in these sample estimates are not appropriate for a specific project, the components of these models may be used to formulate the best estimating method for the job.

System Components

The basic components of a roof are the *deck or substrate, insulation, waterproofing membrane, protective surfacing* (if applicable), *flashing, counterflashing,* and *perimeter termination devices* (Gravel Guard). All of these elements work together to create a watertight envelope for the building interior, as shown in Figure 1.1.

Roofing is most commonly designated in specifications under Division 7, "Thermal & Moisture Protection," using the Construction Specifications Institute's MASTERFORMAT classification system. Division 7 covers building dampproofing, waterproofing, caulking, sheet metal work, and a variety of roofing materials and systems.

The Major Types of Roofing

Most often, the type of roofing system selected for a project is greatly influenced by the knowledge and experience of the designer. In other cases, the owner may have contact with someone in the roofing industry who encourages the use of the latest roofing system. Local fire codes or the community's or owner's architectural guidelines for appearance may also dictate one roofing system over another. The system selected should be one that has demonstrated performance in the area where the project will be constructed. There should also be at least one local qualified roofing contractor who can make any necessary future repairs.

A roofing system consists of many different items made by different manufacturers. There are also related items that must be accounted for which are not roofing materials, but are required for the success of the roofing system. These items include: flashing materials, roof deck insulation, roof hatches, skylights, gutters, and downspouts.

Although the roofing system is a low percentage cost item compared to the square foot cost of the complete building, the system's cost and life expectancy should be carefully evaluated. Water penetration is usually the major problem in building construction and maintenance. If care is taken in the selection of a proper roofing system, much time, money, and aggravation will be avoided later on.

There are four basic categories of roof covering:
- Built-up
- Single-ply
- Metal
- Shingles and tiles

Built-up Roofing

A built-up roof (Figure 1.2) is composed of three elements: felt, bitumen, and surfacing. The felts, which are made of glass, organic or polyester fibers, serve much the same purpose as reinforcing steel in concrete. The felts are necessary as tensile reinforcement to resist the extreme pulling forces in the roofing material. Felts installed in layer fashion also allow more bitumen to be applied to the whole system. Bitumen, either coal-tar pitch or asphalt, is the "glue" that holds the felts together. It is also the waterproofing material in the system.

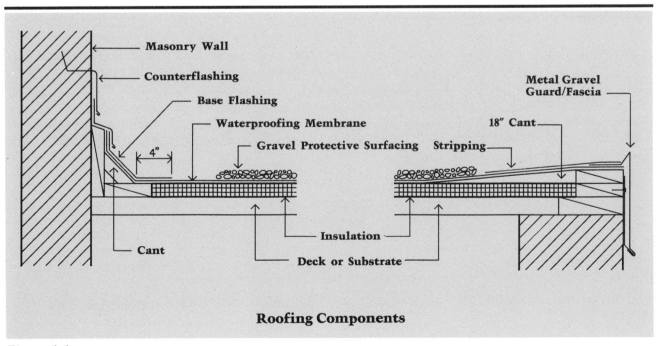

Roofing Components

Figure 1.1

The surfacings normally applied to built-up roofs are smooth gravel or slag, mineral granules, or a mineral-coated cap sheet. Gravel, slag, and mineral granules may be embedded into the still-fluid flood coat. Gravel and slag serve as an excellent wearing surface to protect the membrane from mechanical damage. On some systems, a mineral-coated cap sheet is applied on top of the plies of felt. The mineral-coated cap sheet is nothing more than a thicker or heavier ply of felt with a mineral granule surface.

The most common built-up systems available contain two, three, or four plies of felt with either asphalt bitumen or coal tar pitch. Almost all systems are available for application to either nailable or non-nailable decks. All systems may be applied to rigid deck insulation or directly to the structural deck.

Single-Ply Roofing Systems

Single-ply or elastomeric roofing (Figure 1.3) falls into three categories: thermosetting, thermoplastic, and composites. Single-ply roofing can be applied loose-laid and ballasted, partially adhered, and fully adhered. The loose-laid and ballasted applications are the most economical. They involve fusing or gluing the side and end laps of the membrane to form a continuous nonadhered sheet, held in place with a ballast. Partially adhered, single-ply membrane is attached with a series of strips or plate fasteners to the supporting structure. Because the system allows movement, no ballast is usually required. Fully adhered systems are uniformly, continuously adhered to the manufacturer's approved base.

Figure 1.2

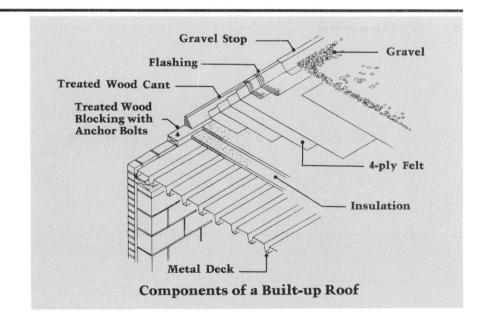

Components of a Built-up Roof

Metal Roofing Systems

Metal roofing systems may be divided into two groups: preformed metal and formed metal. Preformed metal roofs, available in long lengths of varying widths and shapes, are constructed from aluminum, steel, and composition materials, such as fiberglass. Aluminum roofs are generally prepainted or left natural, while steel roofs are usually galvanized or painted.

Preformed metal roofing is installed on pitched roofs according to the manufacturer's recommendation as to minimum required pitch. Lapped ends may be sealed with a preformed sealant (usually available from the manufacturer) to match the deck configuration.

Preformed deck is installed on purlins, or supporting members, and then fastened with self-tapping screws with an attached neoprene washer to prevent leakage. Span lengths depend on roof pitch, deck thickness, configuration, and geographical area.

Formed metal roofing is used on sloped roofs that have been covered with a base material (plywood, concrete, etc.). It is generally chosen for aesthetic rather than economic reasons. Typical materials include copper, lead, and zinc alloy. Flat sheets are joined by tool-formed batten-seam, flat-seam, and standing-seam joints. Solder or adhesive is subsequently applied.

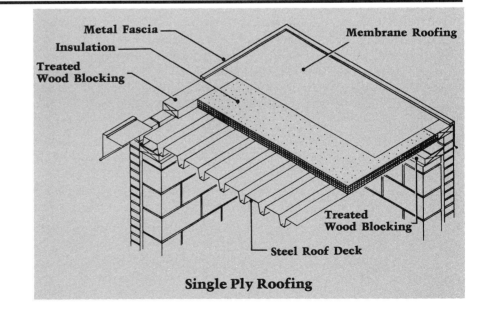

Single Ply Roofing

Figure 1.3

Shingle and Tile Roofs

Shingles and tiles are popular materials for covering sloped roofs. (Figure 1.4) Both are *watershed* materials, which means that they are designed not to retain water, but to direct the water away from the building by means of the sloping or pitching of the roof. Most shingled roofs require a pitch of 3" or more per foot to perform correctly. Shingles may be installed in layer fashion with staggered joints over roofing felt underlayment. Nails or fasteners are concealed by the shingle course above.

Shingle materials include asphalt, wood, metal, and masonry (tile). Asphalt shingles are available in various weights and styles, with three-tab, 240 lb. being the most economical. Wood shingles may be either *shingle* or *shake* grade, with shakes being the more expensive of the two systems, but generally more aesthetically appealing. Metal shingles are either aluminum or steel and can be coated or uncoated.

Slate shingles and tiles of clay and concrete require stronger structural systems to support the added load imposed by the heavier roofing materials. The initial added cost for slate and tile roofs may be offset by the higher replacement and maintenance costs usually associated with other roofing systems.

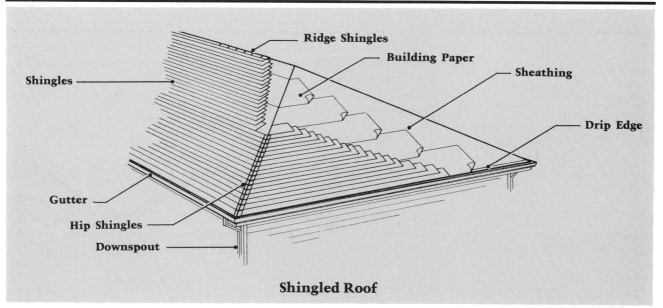

Shingled Roof

Figure 1.4

Summary Designers, owners, facilities managers, and contractors are all faced with new requirements for researching manufacturers' products and coordinating their projects as the variety of available roofing materials and methods grows. Clearly, those who are well organized and informed have the best chance for involvement in successful and profitable building projects. This book is intended to provide a basic working knowledge of roofing components, and guidelines for the requirements of installation, selection, and maintenance. Chapter 2 begins with the fundamental principles for avoiding roofing failures.

CHAPTER TWO

Avoiding Roofing Failures

CHAPTER TWO

AVOIDING ROOFING FAILURES

Design and Job Site General Conditions

In order to ensure an optimal roofing project, no matter what system is selected, both designer and installer must make a habit of incorporating quality control procedures *before, during*, and *after* the roof system is installed. This chapter addresses the components of project supervision and control which can be indispensable to designer and owner, before and during construction. (Post-construction measures are covered in more detail in Chapter 14.)

Pre-Roofing Conference
Conducting a pre-roofing conference is the first step toward a quality roof installation. As recommended by the National Roofing Contractors Association (NRCA), a representative of the owner, architect, general contractor, mechanical contractor(s) and the roofing contractor should all attend. The roofing manufacturer's representative should also be on hand to discuss and/or resolve with the architect any discrepancies between plans and specifications and the recommended installation procedures and details of the roofing material manufacturer. If Factory Mutual (FM) or Underwriters' Laboratories (UL) insurance requirements do not agree with codes, ordinances, or project specs, these discrepancies should be resolved in the presence of all parties at this meeting. The proceedings, conflicts, and conclusions should be noted for the record, and copies sent to all parties.

The Agreement
At a minimum, all parties participating in the pre-roofing conference should agree to the following:
- The work sequence for the deck, rooftop curbs, penetrations, equipment, and roofing material placement.
- Whether or not a temporary roof is required (and if so, who will pay the additional cost).
- The work methods and protective measures needed to prevent roofing damage by other trades.
- The method of approval for changes in roof details and flashings.

Causes versus Symptoms

Splitting and Blistering
In a survey conducted by the NRCA, it was found that the major direct causes of failure in built-up roof systems are *splitting* and *blistering*. These system defects are illustrated in Figure 2.1.

These defects are not *causes* of failures, but are *symptoms*, or, to use an engineering term, *failure modes*. The symptoms result from improper design or material selection, and/or faulty installation practices. Some of these problems are described in the following sections.

Roof Deck Selection

Modern Materials
There has been an increase in splitting and blistering since the design industry began to move away from the use of traditional rigid decks, made of concrete or solid wood, for commercial use. Less rigid decks —such as light gauge sheet metal, "wet" lightweight concrete fill, and shredded fiber panels — now predominate, sometimes compromising the integrity of the roof system. Lightweight metal decks are spot-welded to bar joists, usually on six-foot centers, exhibiting a "trampoline" effect when live loads such as foot traffic, precipitation, and wind uplift are imposed.

Placement on Uncured Concrete
Lightweight concrete "wet decks," when not sufficiently cured, also contribute to the increased failure rate of built-up roofs. This problem occurs when the roofing subcontractor is instructed to begin installation of insulation and roof membrane before the concrete deck has fully dehydrated. This practice causes blisters, commonly referred to as "mole runs," especially in cold weather when moisture condenses in the insulation and migrates to the membrane. When relatively small, a blister formed in this way, without a break in the membrane, imposes no major problems. However, if sufficient moisture is present — if the concrete was

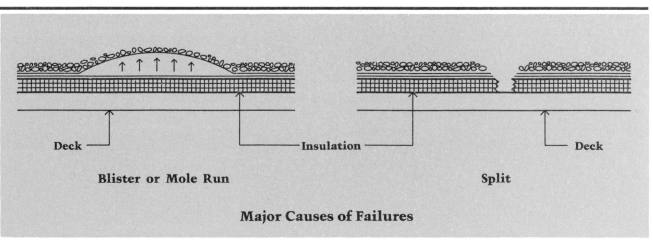

Deck   Insulation   Deck

Blister or Mole Run   **Split**

Major Causes of Failures

Figure 2.1

very "green," or moist when the roof system was installed — heating/cooling cycles create a "bellows" effect, enlarging the blister(s) and causing the membrane to split.

NRCA Deck Dryness Test

Project architects, managers, and general contractors should not compromise the eventual serviceability of the building to meet project deadlines. The roofing contractor must verify that concrete "wet" decks are sufficiently dehydrated, or cured, prior to placement of insulation or membrane. The NRCA has suggested a test for deck dryness which may serve as a guide for contractors and project inspectors. See Appendix C for the NRCA test and related suggestions for checking deck moisture content.

Traffic Damage

Equipment service personnel are not usually aware of the roof membrane's susceptibility to damage, and may transport heavy equipment or materials across the roof without adequately protecting the roof membrane. This traffic, along with dropped tools or equipment, often punctures the membrane or creases it enough to cause an eventual split. These breaks in the membrane create a path by which moisture may enter the insulation below, reducing its "R" value. The introduced moisture can also initiate blistering of the membrane, as moisture becomes trapped between the membrane and the insulation and is transformed into water vapor by solar radiation, creating high vapor pressures.

Protective Walkways

Roof system designers and facilities managers can minimize problems caused by traffic by specifying and/or installing protective walkways, when the roofing is installed. (See Figure 2.2 for one example.) Protective paths or walkways should be located between all roof access points and any rooftop equipment or other areas requiring regular service or access. Each type of roofing requires its own type of protective walkway and installation. The best source of information on the appropriate walkway system can be found in the manufacturer's published installation details and guidelines.

Roof Drainage

"Dead Level" Roofs

It has become a basic principle within the construction industry that flat or "dead level" roof decks are not acceptable design. Manufacturers of roofing materials state, as a general requirement, that roofs should allow for proper drainage. Unfortunately, many design professionals continue to locate drain inlets or scuppers adjacent to columns or on frame lines at the building perimeter. Over time, these areas tend to become the highest points in the roof when the deck deflects under structural and live loads. This deflection leads to ponding on the roof, thereby compounding the problem of insufficient drainage. Vegetation can become established in the ponds, its roots penetrate the roof membrane, and water is allowed to seep into the underlying insulation or deck. This moisture, when combined with solar radiation, causes the asphalt to emulsify, drawing the vital waterproofing oils out of the membrane. This bond break also puts increasing stress on the felts which, due to their relatively low tensile strength, eventually split.

To decrease the possibility and/or severity of ponding, designers

Expansion Joints

and installers must make roof drainage a priority. Drainage is discussed in more detail in Chapter 5.

In most applications, thermal stress caused by the expansion and contraction of large areas of roofing can be severe. In order to retain the integrity of the roof, designers normally use one of two devices: expansion joints or area dividers. If these devices are omitted or improperly installed, the roof system will eventually split or form ridges. Roofing expansion joints should be placed wherever a building expansion joint occurs. The joint should not terminate within the interior portion of the roof, but should continue to the parapet or roof edge. See Figure 2.3 for a diagram of a roof expansion joint.

It is good practice to locate expansion joints wherever there is a change in roof direction, dimension, height, or material. Interior spaces in the same building with vastly different temperatures and/or humidity conditions (such as office/warehouse, office/manufacturing, etc.) should also be separated by an expansion joint.

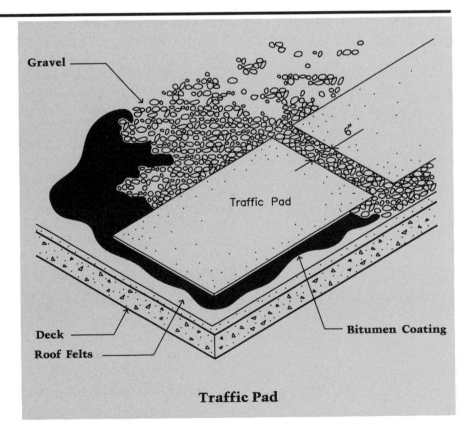

Traffic Pad

Figure 2.2

Area Dividers

If the stress relief required in the membrane is greater than the building expansion joints can absorb, then *area dividers* (shown in Figure 2.4) will be needed. Area dividers are normally spaced at 150- to 200-foot intervals, and located between building expansion joints.

Drainage should be designed and located to allow all areas of the roof to drain as quickly as possible. Area dividers and expansion joints should never be placed so that they impede the flow of water to drains or scuppers. Refer to Chapter 5 for more information on drainage.

Temporary Roofing and Re-roofing

Temporary "Dry-In" Roofing
It is not uncommon for a roofing contractor to be required to install a temporary roof to "dry in" the building, even before walls are up. The first ply or two of waterproofing membrane is installed and mopped off, providing other trades with a dry working environment, and allowing the roofing contractor to perform his work under better weather conditions. The type of material and the number of plies required depend on how moisture-sensitive the interior construction is, and the duration for which temporary protection is required.

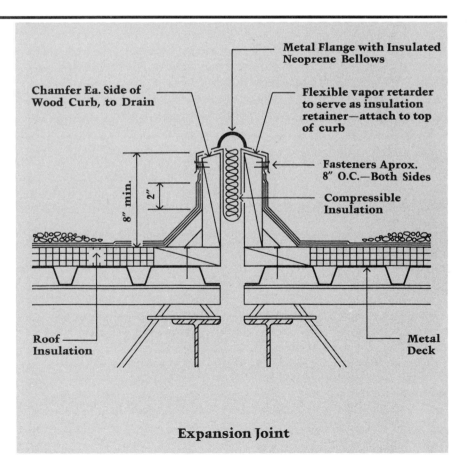

Expansion Joint

Figure 2.3

Surface Preparation

If insulation has been used in the temporary roof, or when re-roofing, a thorough surface examination should be conducted prior to installing the new roofing. After the other trades are finished installing accessories and equipment, the roofing contractor is able to finish the roof system. The existing insulation should be firmly (preferably mechanically) attached to the deck, and all surfaces cleaned and prepared for the new insulation and/or membrane, according to manufacturer's directions. Any damage to the original membrane of an existing roof must be repaired before installing the new system. If proper preparations are not made, moisture intrusion, blistering, and delamination of the roof may occur, often before the general contractor has even completed his work.

Weather Considerations

Precipitation

Roofing materials should never be applied when precipitation is imminent or in progress. Mist, drizzle, rain, snow, or sleet, even when light or of short duration, must be allowed to fully evaporate before the roofing material is applied. When built-up roofing is installed in damp conditions, such as early in the morning before dew has evaporated, water vapor may be trapped in the insulation or between felt plies.

Extreme Cold or Heat

When built-up roofing is installed in very low temperature conditions, the asphalt drops too quickly below the equiviscous temperature range (EVT) as it is applied. The EVT is that range of temperature which allows for optimum dispersion and coverage of the asphalt. The material, when cooled in this way, tends to cause heavy moppings and voids in the membrane. Extremely high

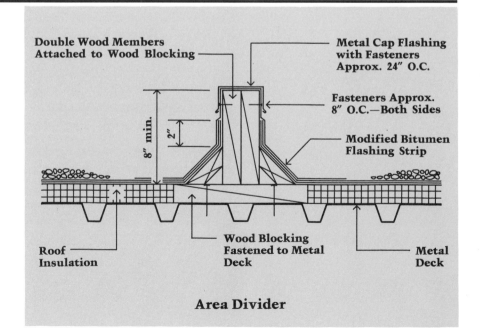

Figure 2.4

18

ambient temperatures during application can also present a problem, since the bitumen will stay in a pliable state longer than usual. This allows potential damage from such sources as foot traffic pulling away the material.

Wind

The combination of low temperature and high winds affects the application temperature and makes uniform application more difficult. High winds may also cause uneven coverage, and lead to the problem of overspray. Applying built-up roofing materials during periods of high wind should be avoided. Installation of roofing materials under any one or a combination of the weather conditions mentioned above can profoundly diminish the quality of the installation, and affect the long-term serviceability of the membrane. The installer, general contractor, and owner would all do well to assure that installation takes place under favorable weather conditions.

New or Untested Materials

Manufacturer's Specifications

A significant cause of past roof failures was the acceptance by the industry of certain new specifications — both those published by manufacturers and those written by project designers. A case in point is the *two-ply specification*, which was touted at one time by several manufacturers as being equal in performance to the traditional four-ply built-up roofing (BUR) method. Designers and industry tradespeople accepted the new specification implicitly. Many roof failures (and lawsuits) later, the claims were proven false, after great expense to contractors, owners, and designers.

Industry Standards

Within the past two years, the American Society for Testing and Materials (ASTM) and others (see Appendix A) have been involved in more comprehensive testing than ever before in the history of the roofing industry. It is hoped that, very shortly, a consensus will be reached within the industry on standards for materials, methods, and specifications. In light of the great volume of new products and application methods coming into the market today, the specifier and/or installer is well advised to regularly monitor the latest publications and standards offered by ASTM and the National Roofing Contractors Association (NRCA).

Roof System Selection

Inappropriate System Type

In many cases, the cause of roof failure may not be a faulty system specification *per se*, but may be due to the fact that the system selected is not appropriate for a particular application. Examples of inappropriate selections are: the use of shingle roofing on a roof, the slope of which is less than three-in-twelve; or the use of BUR or single-ply on a steeply-pitched roof.

Preliminary Considerations

The specifier must first consider conditions under which the roof must operate, such as:

- Normal ambient temperature and humidity range
- Building interior temperature and humidity range
- Rainfall and other precipitation
- Abrasion, expansion/contraction from ice and snow
- Wind uplift

- Expected foot traffic
- Adjacent vegetation or trees
- Birds, rodents, and other pests (see the next section)
- Airborne pollutants or adjacent noxious activity
- Size and configuration of roof deck
- Parapet and/or intersecting wall details
- Other building components requiring roof interface
- Rooftop equipment requirements

Also important in the selection process are the costs, aesthetics, and maintainability of the system chosen. Rapidity of installation may be a factor due to project time constraints. Developing a *Roof Selection Chart* (See Chapter 17, Figure 17.1) can give the specifier a relatively quick overview of system characteristics.

Rooftop Lighting and Insect Damage

In March, 1988 a professional roofing magazine article described a bizarre but apparently avoidable, phenomenon. In about a dozen documented cases ranging from Washington State to Florida, beetles have bored through roof membranes, causing leaks. It was determined that the beetles are attracted to lights (especially mercury vapor) mounted on, over, or near roof surfaces, including nearby billboard lighting. Falling to the roof, they burrow into the roof substrate, seeking protection from the sun during the day. The types of roof membrane affected were asphaltic BUR, modified bitumen, and single-ply roofing. Evidently, no instance has yet been found among coal tar BUR. It is advisable to exercise care in the selection of roof membranes where billboards may exist adjacent to a planned roof installation, or when rooftop lighting is required. The roof specifier should discuss the types of luminaire to be used with project electrical engineers before mercury vapor fixtures are specified.

Incompatible Materials

Another cause of roof failure is a chemical incompatibility of materials specified for use in the same system. For example, numerous failures have resulted from the use of coal tar pitch bitumen with asphalt-saturated glass fiber felts. This combination still appears in at least one manufacturer's specification. Inspection cuts made through the membrane of roofs containing this combination have revealed that the coal tar pitch has migrated out of the felt plies containing the chemically incompatible saturant, allowing moisture into the system, and causing eventual failure. The membrane material may also be incompatible with the insulation, adhesive, or bonding substances chosen. In time, the combination of the two materials can lead to a fracture of the bond between the materials, delamination, or even tearing of the material itself. It should be standard procedure to check all materials selected within a system for chemical compatibility.

Roof Venting

NRCA Criteria

Many roof system materials such as insulation and decking should be vented to relieve vapor pressures that build up within them. The NRCA has published the following venting recommendations, based on actual field experience. These regulations are considered to be a minimum list of venting criteria:

1. All closed plenum, unvented roof systems should be vented by means of moisture relief vents.
2. Where the minimum dimension of a building is 50 ft. or less, *edge venting* may be used (open gravel guard or fascia).

3. Otherwise, *edge* and *field* venting are required.
4. If the average relative humidity is high, one-way vents should be used for the *field* vents (those vents in the middle area of the roof).
5. Vent stacks in the field portion of the roof should be installed at the rate of one vent per 1000 square feet or less of roof area. A minimum of two vents should be used, regardless of size.

See Appendix D for further details on venting procedures.

Roof Flashing

Proper flashing is one of the most commonly overlooked aspects of the roofing design/construction process, yet it is a common cause of roof failures.

One problem is that accessories, equipment bases, and curbs are often not high enough to accommodate proper flashing. Ductwork installed on the roof may also be poorly supported or inadequately flashed.

The project architect and mechanical/electrical engineers should carefully coordinate equipment dimensions and curb details, to avoid forcing the contractor to field-modify unsuitable details. These areas of coordination, as mentioned at the outset of this chapter, should be documented at the pre-construction roofing conference. This advance planning not only resolves potential liability problems before they occur, but also expedites the project and encourages cooperation between tradespeople. See Chapter 5 for detailed coverage of flashing.

Summary

Proper roof installation takes into account all of the possible causes of roofing failures, and incorporates the appropriate preventive measures. This means establishing and carrying out quality control procedures before, during, and after the system is installed. This task involves research and coordination in the design and preconstruction stages, and follow-up of proper methods and materials during installation and the later maintenance programs. The responsibilities of preventing roofing failures are clearly wide-ranging and long-term, but they are also the most crucial to the roofing contractor, designer, and owner.

CHAPTER THREE

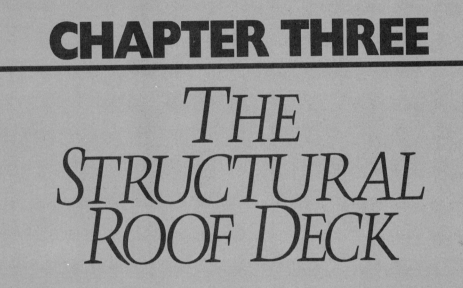

CHAPTER THREE

THE STRUCTURAL ROOF DECK

The structural roof deck, as its name implies, is an integral part of the building structure. As such, it supports not only the roofing system, consisting of insulation, membrane, and surfacing, but also live loads, such as wind and various forms of precipitation. In addition, economy of design often dictates that the deck perform beyond the simple function of carrying the roofing and gravity loading. The deck also acts as lateral bracing for compressive sections of slender framing, such as joists or purlins. Additionally, the structural deck may be used as a diaphragm to receive, distribute, and deliver lateral forces to walls or other buttressing elements of the building.

All roof decks must be compatible not only with their accompanying system components, but also with the service environment. To successfully support a roofing system, the deck must have the necessary strength, stiffness, dimensional stability, and durability to provide a good foundation for that system.

During the design period, decisions are made that will affect both the building's function and its maintenance costs for the life of the building. Decisions regarding the selection and design of the deck may provide versatility, or severely limit the choices of roofing and options for future re-roofing. The deck should, therefore, be designed by someone with a thorough understanding of roofing. If the building designer is not an expert on roofing technology, a professional should be consulted who can match the roofing system to the structure. Decisions made in order to reduce the initial cost of the deck may actually cause great expense over the long term, due to abnormally frequent maintenance requirements or premature replacement of the roofing.

Deck Classification

Decking as a structural component of a building is subject to the same design parameters as other structural components: it must be safe, functional, and economical — in that order. Cost is always a consideration, but economy must never infringe upon reasonable and necessary standards of safety and function.

Decks are classified in a number of different ways, based on different criteria. Codes or insurance-related agencies classify decks as *combustible* or *noncombustible*; built-up roofing manufacturers and their specifiers classify them as *nailable* or *non-nailable.* Structural engineers and architects usually classify decks by material, such as *concrete, cementitious wood fiber, steel,* or *wood.* In this chapter, decks are categorized by material, with some additional discussion about applicable geometry and special physical properties.

Considering the wide variety of types of decks and roofing systems, the number of possible combinations is staggering. As some roof system components may be incompatible with certain deck types, the designer or roof specifier should be aware of potential problems with certain combinations in advance. This chapter covers the characteristics of each of the various types of decks and the potential problems that these arrangements may present to the roof system. Such problems can be avoided through optimal system selection and design.

Steel Decking

Although there are local regions where certain types of decks predominate (such as concrete in hurricane-prone coastal areas, or wood in the Pacific Northwest), steel is used for the majority of nonresidential roof decking in the United States. Steel decking on open web steel joists is the most commonly encountered substrate for both BUR (built-up roofing) and single-ply systems (see Figure 3.1). In these systems, rigid board insulation is most commonly used.

Physical Properties
Throughout the years, many sizes and shapes of steel roof decking have been manufactured. The industry has, through efforts of the Steel Deck Institute (SDI), reasonably standardized the geometry

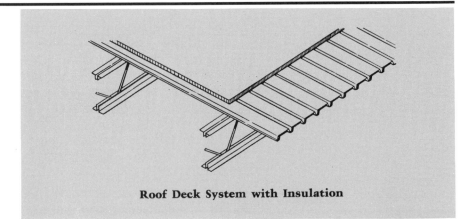

Roof Deck System with Insulation

Figure 3.1

and specifications of cold-rolled roof deck. Steel decks complying with SDI specifications can be obtained from member companies in 1-1/2", 2", 3", 4-1/2", 6", and 7-1/2" depths. Common rib spacings are 6", 7-1/2", 8", 9", and 12" on center. These decks are available with and without stiffening elements. Panel widths vary from one to six rib spaces. The most commonly used steel decking is 1-1/2" deep, with 6" rib spacing, in panels 24", 30" or 36" wide.

Common thicknesses of steel roof decking are 22 gauge (0.0299"), 20 gauge (0.0359"), and 18 gauge (0.0478"). SDI specifications permit minimum thicknesses before coating of 0.028", 0.034", and 0.045", respectively, for these three gauges.

Current designations of 1-1/2" deep decking are NR (narrow), IR (intermediate rib), and WR (wide rib). Figure 3.2 shows SDI standard rib opening control dimensions.

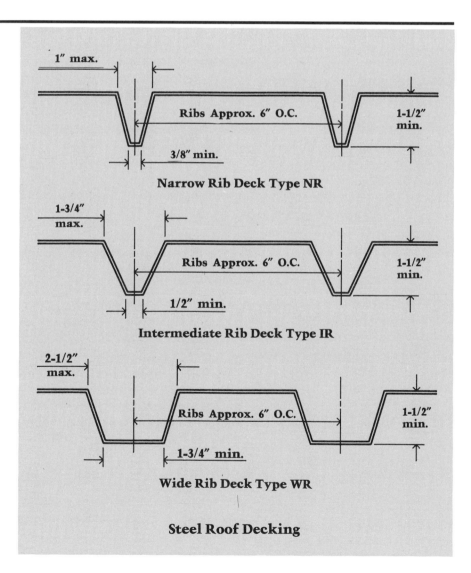

Narrow Rib Deck Type NR

Intermediate Rib Deck Type IR

Wide Rib Deck Type WR

Steel Roof Decking

Figure 3.2

The geometry of the decking cross section has important effects on economy, strength, and stiffness. Unfortunately, these effects are not always fully understood by specifiers. As the rib opening widens, the bottom flange widens, increasing the mathematical section properties which define strength and stiffness.

Steel is sold by the pound, and deck panels of the same gauge weigh essentially the same amount, whether they are NR, IR, or WR. However, there are radical differences in their strength and stiffness. Using one national manufacturer's published section properties, progressing from NR to WR increases stiffness by 54% and increases strength by an average of about 92%.

If there were no other considerations, WR section material would always be the most economical: it would permit longer deck spans and consequently, fewer joists or other supporting members. If the same joist spacing is used, wider ribs increase load carrying capacity and decrease deflections under loading. Wide rib sections nest better and permit easier access for attachment than intermediate or narrow rib sections. The benefits of using wider ribs are, however, offset by the fact that thicker and stronger insulation is required to span a wider rib opening without fracturing due to foot traffic or construction loading.

Prior to the "oil crisis" in the early 1970's, rigid insulation board in $1/2''$ or $3/4''$ thickness was commonly used over NR deck in moderate climates. Narrow ribs were necessary for use with thin insulation, and the large top flange area was desirable when insulation was being mopped down. Both of those factors have changed, however. "R" values currently required are rarely less than 10, and insulation is normally no longer mopped to steel decks. Thus, there is little justification for using NR deck. Aside from its inferior physical properties, NR deck nests poorly for shipment, requires swaged ends for proper fit at end laps, and yields a high incidence of faulty attachments because of difficult access. The characteristics of narrow ribbed deck make piecemeal repairs during re-roofing extremely difficult. Repair requires nesting of new material to old at end laps and side laps, when replacing or superimposing a panel.

Finishes

Steel roof decks, even though they are out of the weather, are nevertheless installed in relatively hostile environments and, therefore, must have factory-applied protective finishes. Two finishes are commonly used: *paint* and *galvanizing*. Paint is the cheaper and thus more predominant finish, but it has certain drawbacks. Using this method, primer is applied to the flat coil stock before the ribbed panels are rolled from it. However, if the underside of the deck is painted after installation, its surface is subject to continued rusting. Any entrapped or migrating moisture due to vapor or leaks can seriously damage inadequately protected steel deck. Another problem is deformed and cracked primer film at the newly formed "corners." Most painted decking exhibits rust along these bends even before it is in place in the structure. The other alternative, galvanized decking, costs only a few cents more per square foot than factory-primed decking, yet a galvanized finish is far more durable and protective.

Both painted and galvanized decking suffer finish damage when they are attached by welding. The welds and other affected finish should be touched up with compatible material: rust-inhibitive paint for a painted deck, and either zinc chromate or liquid galvanizing for galvanized deck.

Attaching Steel Deck to the Structure

Steel decking must be properly attached to the supporting structure to permit its proper function as a substrate for the roofing system. If it rolls, flutters, or deflects excessively at its side laps, the roofing system may be damaged.

Common means of attachment are welds, screws, or power-driven studs. These methods are discussed in more detail in the following sections.

Welding: Welding is by far the most common attachment method because it is generally thought to be the most economical. However, it can actually be more costly than screw attachment, when it is properly specified and carefully inspected by a knowledgeable inspector. These measures require additional time and manpower. If the workmen attaching the decking use poor technique, excessive amperage, or the wrong type of welding rods, the result is often damage to the joists, or a weak or nonexistent attachment of the deck.

When welding, it is good practice to inspect and mark each attachment line of steel roof deck before installing the roofing system. Often 25 to 50 percent of the welds must be reworked. At end laps, and particularly at end/side laps where four thicknesses of material are stacked, the welder may fuse the sheets together, but not to the supporting joist. Industry-accepted specifications permit welding without weld washers or clips on 22-gauge and heavier metal, but experience has proven that 22-gauge metal is too thin to accommodate an adequate weld to the joist. Weld washers should not be omitted unless the material is 20-gauge or heavier.

Use of high amperage when welding thin material leads to poor weld configuration and damage to joists by "blowholes." If the gap between the welding rod and the decking is too great, a "hot arc" results, with very little transfer of weld metal. Welders commonly use E6011 electrodes —deep penetration, all purpose, "quick freeze" type. This is a poor choice for making welds of thin sheet metal to steel framing members, nor is it appropriate for thin, high strength steel joists. Many joists have top chord members constructed of of $1/8''$ thick angles which are easily damaged. Welding rods used for deck installation should instead use a lower amperage range and have a fast deposition rate. There is no need for full penetration weld techniques for deck installation (as would be used for structural welding). Quick freeze characteristics are also unnecessary, since all deck welding is done on a flat surface, not overhead.

Screws: The initial cost of self-drilling screws may seem higher than welding in terms of material and labor. However, this method results in a neater installation than welding, without damage to deck coating or joists. If the costs of extra inspection, reworking, touch-up painting, and long-term rust potential of the welded attachment method are taken into consideration, screw attachment is generally found to be a better choice.

Technique: Particular care must be exercised not only in attaching the deck to the structure, but also in joining deck panels at side laps. These are vulnerable locations for differential deflections under concentrated loads (foot traffic, wheels, and the like). SDI specs require side lap fasteners if the deck span exceeds five feet, with the fastener spacing not exceeding three feet. Factory Mutual (in FM 1-28 *Loss Prevention Data, Insulated Steel Deck*, rev. 8/83) recommends that "Mechanical fastening should be provided at all deck side laps. Spacing between side lap fasteners and bar joists or beams should not exceed 3 ft. (0.91 m)." In other words, side lap fasteners are required when the joist spacing exceeds three feet. This requirement is more stringent; it results in a stiffening of the deck and protection of the insulation and roofing membrane.

Concrete Decks

Concrete decks make excellent substrates for roofing systems with a first ply, or layer, which is hot mopped in place. The initial cost is usually higher for concrete decks, as compared to other types of decks. However, concrete offers two distinct advantages.

- **Stiffness:** Concrete displays no local distortions due to concentrated loading, and displays extreme resistance to flutter and lifting in high winds.
- **Durability:** Concrete does not decay, is fire-resistant, and is relatively resistant to water damage.

Structural Concrete

Structural concrete decks may be *cast-in-place* or *precast*. Cast-in-place concrete decks are either *conventionally reinforced* or *post-tensioned*. Precast decks are also either conventionally reinforced (commonly designated "reinforced") or pre-stressed, but pre-stressed is by far more commonly used.

Concrete decks are superior to lightweight decks for ballasted roofing systems. For insulated or mechanically attached single-ply systems, however, the cost of attaching insulation to the deck becomes a factor. When not topped with cast-in-place concrete or underlayment, the insulation board may fracture or membranes may split. Adjacent deck sections must be properly attached structurally in order to prevent differential vertical movement. It is also necessary to level the butted ends of sections if they are out of level. If lateral movements are possible in the deck system, the roofing membrane must have adequate flexibility in order to prevent splitting.

Pre-stressed members deflect a greater distance than cast-in-place reinforced structural slabs under comparable loads. They also gain and lose camber from heating and cooling of the upper surface, if not well insulated. Considerable movement can also take place between the ends of members in multiple span arrangements; this movement must be addressed in the selection of the roofing system, its attachment, and flashings. Flashing along terminal (side) edges of long members must often accommodate several inches of vertical movement between the deck and the parapet.

In lightweight insulating concrete decks, venting is not provided beneath the concrete. Thus, efforts should be made to assure proper drying of the overlay prior to roofing. (See the Deck Dryness Test in Appendix C.) Attention should also be paid to lateral venting of moisture vapor between the concrete and the underside of a built-up roof (BUR) membrane. This may be accomplished with a *venting base sheet*; one-way *roof vents* may also be helpful. (see Appendix D).

Hot mopping of BUR membranes to lightweight fills is discouraged by roofing authorities. Instead, special fasteners are recommended to attach base sheets to this nailable substrate. Such fasteners usually have a two-piece shank and an oversized head. The shanks mechanically anchor into the media by expanding, deforming, or barbing as they are driven. The plain shanks of standard roofing nails do not offer sufficient resistance to withdrawal. To mechanically attach insulation boards to lightweight fill, special fasteners are available with large, coarse threads and integral, or two-piece, disc heads. Some also have internal "barbs" that are actuated from the top after setting the screw. These special screws are spaced more widely than nails (up to three times more widely spaced), for attaching a base sheet.

The larger value of uplift resistance required for the insulation fasteners is often beyond the pull-out strength of lightweight insulating concrete. Uplift tests should be run, using the planned fastener in the lightweight concrete on-site, to confirm that the required design values can be achieved. Otherwise, alternative methods of attachment must be used.

Gypsum Concrete

Gypsum concrete decks are sometimes used as structural decks for short spans between purlins or sub-purlins when their particular properties are desirable or acceptable. Compared to structural concrete (Portland cement concrete), gypsum concrete is lighter and cheaper, is nailable, and has modest insulating and good sound-deadening properties. The drawbacks of gypsum concrete are that it is comparatively weak in its best condition, and unreliable when wet or subjected to prolonged high humidity. Furthermore, it cannot be structurally analyzed like structural concrete.

Gypsum is used over formboard, reinforced with galvanized, octagonal wire mesh ("chicken wire" or "plaster netting") for structural deck, spanning between truss tees used as sub-purlins (shown in Figure 3.3). When used in this manner, the gypsum provides a nailable substrate, and has base sheets fastened with special nails similar to lightweight insulating concrete. Gypsum concrete is denser and stronger than lightweights using perlite or vermiculite aggregates, but it can become weak when wet. It is also heavier, weighing 38 to 50 pounds per cubic foot, depending upon the type of filler used in the mix.

Manufactured precast gypsum planks, cured and dry when installed, provide a ready roofing base. They usually include metal tongue-and-groove interlocking edges for leveling and load distribution. However, stabilizing this type of decking and providing bracing for the purlins supporting them may require some difficult connections.

Wood Decks

Wood is versatile, easily obtained, and economical. It can be nailed, screwed, sawed, or drilled with relative ease. Local repairs are more readily accomplished using wood than is the case with other decking materials. Wood has the strength to sustain a high degree of overload for short periods without failure. This quality is reflected in codes and design standards that permit increases in allowable design stresses for short duration loading: 100 percent stress increases for impact, 33 percent for wind, or 15 percent for snow loading.

Ventilation Requirements

Considering that wood is used as the decking for nearly all residential roofs and represents a moderate percentage of decking in other construction categories, it is the most common of all deck materials. Nevertheless, it continues to be misapplied and abused. A prominent example of improper use of wood decking is failure to ventilate. Since the roof offers the greatest potential for thermal transfer in a low-rise building, temperature and humidity differences create "vapor drives" through many ceiling and roof structure materials. Wood roof framing, if not pressure-treated against decay, *must be ventilated*. Air, wood, and decay-propagating organisms are ever present; suitable decay-producing temperatures are anywhere between 45 and 110 degrees F. To prevent decay, wood moisture content should be kept below 20 percent, or the wood should be pressure-treated to make it toxic, or unacceptable, to fungi.

If decay begins, it can be arrested by removing one or more of the required ingredients. However, it cannot be reversed. Decay has a cumulative effect over several years. It may cause weakening, loss

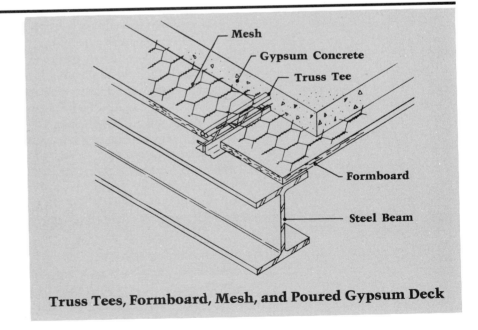

Truss Tees, Formboard, Mesh, and Poured Gypsum Deck

Figure 3.3

of stiffness and elasticity, and reduced holding power on fasteners from decay.

Wood is relatively free of thermal movement when compared to steel or concrete, but it swells or shrinks with moisture changes. In service, plain sawn pine or fir may shrink roughly 2.5% across its face dimension if dried from its marketed moisture content (MC) of 19%, down to an equilibrium MC of 9%. That amounts to $3/16''$ across the face of a nominal 8″ board. The moisture content of the wood varies seasonally when it is subject to changing ambient conditions. In addition to structural strength and stiffness, conditions of use, ventilation, edge spacing, fastenings, and attachment of the overlying roof system components should all be properly planned to achieve optimum service.

Cementitious Wood Fiber Decks

Cementitious wood fiber decks are often used in structures that do not have a suspended ceiling. They are frequently used in gymnasiums and schools. The planks or panels are plant-produced, using treated wood fibers and a binder of Portland cement or gypsum. They are molded with controlled pressure and temperature to modular widths. When used with bulb tees, the planks have rabbeted edges to lay on bottom flanges of the tees (see Figure 3.4). The tees are usually spaced at two feet, nine inches on center and the planks span transversely between them. When used to span longitudinally between joists or purlins, the planks are generally eight feet long and have tongue and groove edges.

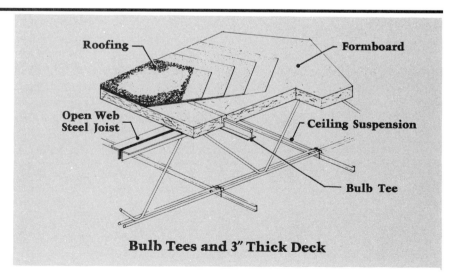

Bulb Tees and 3″ Thick Deck

Figure 3.4

Cementitious wood fiber is multi-purpose; it has the structural capacity to serve as a roof deck, spanning supports up to four feet apart; it is nailable; and it has modest insulating and acoustical values. However, like other multi-purpose materials, it is not without certain drawbacks. In general, the physical properties necessary for one function tend to detract from the optimum performance of another (i.e., the density required for strength detracts from the material's insulating value). With shredded wood as a primary component, cementitious wood fiber is subject to moisture absorption. Shrinking and swelling of the wood fibers results from moisture content changes. If pronounced, these dimensional changes disrupt the cement bindings and weaken the deck. Planks containing gypsum binder are more affected by moisture than those containing Portland cement. Excessively humid conditions can lead to abnormal creep (continued deformation without increase in stress) and result in sagging of the planks.

Summary

The roof specifier, when consulting with the structural designer to arrive at compatible systems, should be aware that the structural deck is the foundation upon which the roofing system is built. Roofing system selections are directly affected and limited by structural decks, and vice versa. The following basic principles should be kept in mind.

- In new design, provide a deck and its supporting structure with the strength and stability to accommodate loading for a variety of roofing systems, attachment of components, and temporary loading during construction. Re-roofing will inevitably be required in the future.
- Provide slope for positive drainage, using a minimum of $1/4$ inch per foot. (See Chapters 2 and 5 for more detail on drainage.)
- Check special loading conditions, such as concentrated equipment loads, snow slides or banks at base of walls of high roof sections, and ponding behind potential ice dam locations.
- Check roof load bearing capacity for any roof that will have more dead load added by re-roofing. Steel and wood structures are more likely to be overloaded by the added weight, but do not overlook potential problems in reinforced or pre-stressed concrete. Long span "flat roof" members with perimeter drainage can be especially troublesome if they have lost their initial camber.

34

CHAPTER FOUR

INSULATION AND VAPOR RETARDERS

CHAPTER FOUR

INSULATION AND VAPOR RETARDERS

Insulation

Roof insulation serves a dual purpose: to reduce heat transfer, and to act as a stable, uniform substrate for the roof membrane. In most cases, roof insulation is installed *between the structural deck and the roof membrane.* Figure 4.1a is an illustration of a conventional roof insulation system. One exception, the Inverted Roof Membrane Assembly (IRMA), is used in severely cold climates. In this arrangement, the insulation rests above the waterproofing membrane. This system, also known as Protected Membrane Roofing (PMR), is shown in Figure 4.1b.

There are many types and thicknesses of material available for roof insulation. Figure 4.2 is a chart comparing the characteristics of various roof insulation materials. The most common of these materials are listed below.

- Polyisocyanurate Foam Board
- Polyurethane Foam Board
- Polystyrene Foam Board
- Cellular Glass Board
- Mineral Board (Perlite)
- Phenolic Foam Board
- Wood Fiber Board
- Glass Fiber Board

   Note: Composite board is a combination of two of the boards listed, with a facing of a different material. The two boards are laminated in the factory to produce a product with the advantages of both materials.

Polyisocyanurate Foam Board
Polyisocyanurate board is a dimensionally stable, closed-cell foam which normally has a glass fiber facing bonded to it in order to receive hot moppings or adhesive. Because of its high thermal efficiency (R = 7.2/in.) and relatively low cost, this material has captured about one third of the market.

There have been some problems with polyisocyanurate board. The most publicized is thermal drift of the "R" value as the insulation ages. There have also been incidences of facer separation. It is best to use materials from the same supplier whenever practical, and to assure that the board manufacturer also cross-references, or approves, the bitumen and adhesive(s) used.

Polyurethane Foam Board

Polyurethane and urethane foam boards have been in common use for a longer period than polyisocyanurate, and were the efficiency "leaders" before the advent of phenolic and polyisocyanurate. Urethane has had the same "thermal drift" problems as polyisocyanurate, but overall, it is still one of the most efficient insulation boards, even if its aged "R" value is applied.

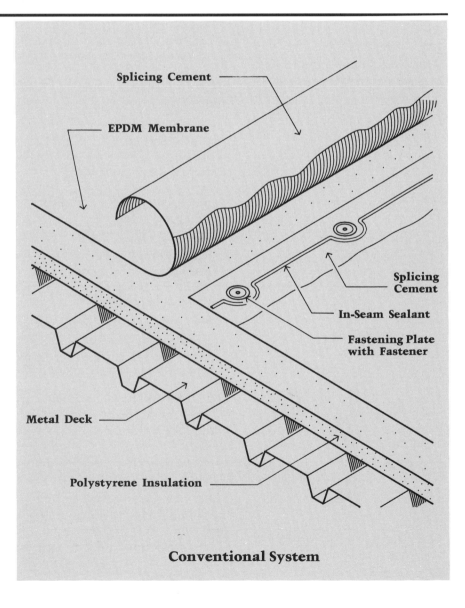

Conventional System

Figure 4.1a

Polystyrene Foam Board

Expanded polystyrene: Expanded polystyrene (EPS) is a polymer (plastic) impregnated with a foaming agent which, when exposed to heat, creates a uniform structure resistant to moisture penetration. Polystyrene has an "R"value of 3.85-4.76 per inch and a compressive strength of 21-27 psi (pounds per square inch) at 1.5 pounds per cubic foot of density. As a comparison, the compressive strength of mineral board (perlite) is 35 psi.

Most expanded polystyrene is used under loose-laid, ballasted single-ply systems. The type of expanded polystyrene often used in these cases is a material with a density of one pound per cubic foot, with an average "R"value of 4.2 per inch. This type of polystyrene is similar to that used in the manufacture of inexpensive picnic coolers.

Using mechanized gravel buggies for ballast installation tends to damage this lighter density EPS. The lighter material is also more likely to "shuffle"or crack under the membrane. The industry now

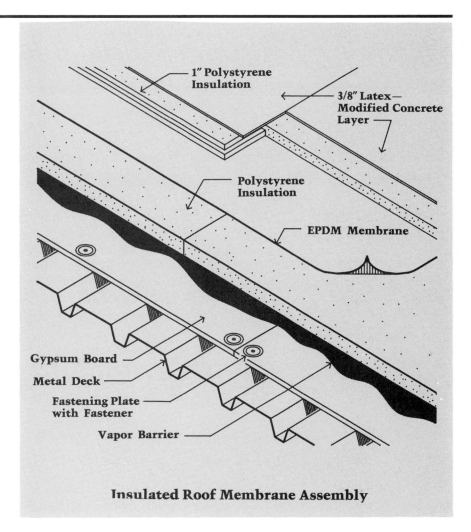

Insulated Roof Membrane Assembly

Figure 4.1b

Characteristics of Various Roof Insulation Materials

Characteristics	TYPE OF INSULATING BOARD								
	Polyisocyanurate Foam	Polyurethane Foam	Extruded Polystyrene	Molded Polystyrene	Cellular Glass	Mineral Fiber	Phenolic Foam	Wood Fiber	Glass Fiber
Impact resistant	G	G	G	F	G	E	G	E	F
Moisture resistant	E	G	E	G	E	G	E		
Fire Resistant	E				E	E	E		E
Compatible with bitumens	E	G	F	F	E	E	E	E	G
Durable	E	E	E	E	E	G	F	E	E
Stable "k" value			E	E	E	E		E	E
Dimensionally stable	E	E	E	E	E	E	E	E	G
High thermal resistance	E	E	E	G	F	F	E	F	G
Available tapered slabs	Y	Y	Y	Y	Y	Y	Y	Y	Y
"R" value per in. thickness*	7.20	6.25	4.76	3.85-4.35	2.86	2.78	8.30	1.75-2.00	4.00
Thicknesses available	1"-3"	1"-4"	1"-3½"	½"-24"	1½"-4"	¾"-3"	1"-4"	1"-3"	¾"-2½"
Density (lb./ft.³)	2.0	1.5	1.8-3.5	1.0-2.0	8.5	16-17	1.5	22-27	49
Remarks	Prone to "thermal drift"	Prone to "thermal drift" Note "A": Should be overlaid with a thin layer of wood fiber, glass fiber or perlite board, with staggered joints.	Somewhat sensitive to hot bitumen & adhesive vapors Note "A": Should be overlaid with a thin layer of wood fiber, glass fiber or perlite board, with staggered joints.	Somewhat sensitive to hot bitumen & adhesive vapors Note "A": Should be overlaid with a thin layer of wood fiber, glass fiber or perlite board, with staggered joints.			Prone to "Thermal drift" relatively new & untested.	Expands with moisture —holds moisture.	Prone to damage from moisture infiltration.

E = EXCELLENT
G = GOOD
F = FAIR

*from ASHRAE 1985 Fundamentals Manual

Figure 4.2

favors using a board with a minimum density of 1.5 pounds of the extruded material, which is far more durable.

Extruded polystyrene: Extruded polystyrene is a closed-cell foam with a low capacity for water absorption (.06% by volume), which makes it ideal for use in the insulated roof membrane assembly (IRMA). Its "R" value is approximately 4.8 per inch.

Installation: Most building codes require the installation of a $1/2$ inch fire-rated gypsum board underlayment when polystyrene is used over a steel deck. The NRCA also recommends that, when using polystyrene board, a thin (minimum $1/4$ inch) layer of perlite or fiberboard "recover board" be overlaid, with joints staggered from the insulation board joints below. Polystyrene board must also be protected from ultraviolet light and it should not be used on buildings above 75 feet in height, unless it is mechanically attached, due to its light weight.

Cellular Glass Board
Cellular glass board is no longer used as often as it once was for roof insulation. Its lower "R" value per inch and relatively high density (thus, weight) has rendered it less popular than the newer materials. It is, however, a moisture-resistant, stable, and durable material.

Mineral Board
Mineral board is composed of expanded perlite, cellulose binders, and waterproofing agents. Low water absorption, stability, superior fire-resistance, and the best compressive resistance of all the insulations currently used, make this material extremely popular. The biggest drawback of mineral board is its relatively low "R" value (2.78 per inch). It becomes impractical and expensive to use four inches of this material to achieve an "R" value of 10, when the same value can be attained using only 1-$1/2$ inches of foam insulation.

When mineral board is used as the base layer of a two-layer insulation system, it must be mechanically attached to the deck. A vapor retarder should then be installed or mopped over the mineral board, followed by a high performance insulation. This system ensures secure attachment to the deck and positive control of vapor, which would otherwise tend to be driven into the insulation during the winter months. By installing two separate layers of insulation with staggered joints, thermal bridging of the fasteners is eliminated, as is bridging at joints. Placing the vapor retarder above the fasteners and maintaining a temperature at the fasteners above the dew point reduces corrosion of the fasteners and the deck.

Phenolic Foam Board
Phenolic foam is relatively new to the industry and is becoming very popular with specifiers because of its "R" value of 8.3 per inch and its competitive pricing. It is a fire-resistant, dimensionally stable, foamed plastic. However, it has not been in the field long enough for its performance characteristics to be proven. It is friable (subject to crumbling), and will fracture if abused. There have been instances where phenolic foam "dished" (raised at the corners) under single-ply membranes. If soaked, phenolic foam will absorb

moisture and lose much of its thermal efficiency. Although not fully tested, it is reasonable to assume that phenolic foam is affected by *thermal drift*, as are the other foams that use fluorocarbon blowing agents. (Thermal drift is the reduction in "R" value which occurs when the fluorocarbon used as the foaming agent in manufacture vacates the material over time, and is replaced with atmospheric air, which has a lower "R" value.)

Wood Fiber Board

Wood fiber board is used as a combination decking and insulation board for many roof systems. It is stronger and more durable than the other board insulation products, and makes a far better attachment substrate. However, these attributes are offset by its relatively low thermal resistance. Many specifiers use a composite board or a combination of wood fiber board and one of the high efficiency foams to create a solid, efficient system upon which to overlay a roof membrane.

Glass Fiber Board

Glass fiber insulation board is comprised of glass fibers, bound by a resinous binder and rolled into rigid board. A top surface (*facing*) of asphalt-adhered kraft paper or foil is applied as protection. This type of insulation is very efficient (R = 4 per inch) and has been tested by both Factory Mutual and Underwriters' Laboratory. This board is relatively soft and prone to moisture, which attacks the binder and causes the material to collapse.

Associated Insulation Materials

Recover Boards

When installing single-ply membranes over hard-surfaced insulation board, fasteners or gravel can work their way up through the membrane, causing leaks that are very difficult to locate. Some specifications call for the use of *recover boards*, made of pressed fibers, over the insulation board. Recover boards are susceptible to moisture contamination and, therefore, should not be used over wet decks.

Vapor Retarders

Vapor retarders used to be referred to as "vapor barriers," until the industry conceded that no true "barrier" may be devised against moisture intrusion using readily available sheet materials or mopped membranes. Every vapor retarder allows *some* moisture to permeate. Moisture permeability through a vapor retarder is measured in *perms*.

A vapor retarder is strongly recommended when the average outside January temperature is below 40 degrees and the indoor relative humidity in winter is 45 percent or above. The NRCA recommends that when a vapor retarder is specified, moisture relief venting must be installed as a means of allowing any trapped moisture to escape from the system. There is, however, strong evidence that roof vents do not significantly contribute to moisture removal in insulation that is already wet. The primary source of moisture in roof insulation is from breaks in the membrane. Therefore, it stands to reason that adding many roof vents in the system will only increase the chances for a leak.

Vapor retarders *should not be used* unless dew point vapor flow calculations clearly indicate that these devices are necessary, or unless the previously stated January temperature and humidity ranges prevail in that location. For most areas of the U.S., the "downward drying" that occurs in the building space during warm seasons of the year will far exceed the "upward wetting" which occurs in the colder periods. There is one clear exception to this rule: in wet-process applications, such as laundries, canneries, and swimming pools, the "upward wetting" potential virtually demands a vapor retarder to restrict contamination of the insulation.

It is very important that vapor retarders, when used, be as homogeneous as possible, with as few seams or breaks as practical.

Summary

Roof insulation is a key element in the properly functioning roof system. It has a dual purpose — reducing heat transfer and providing a stable, uniform substrate for the roof membrane. There are a wide variety of insulation materials available, making research and comparison a must for the roofing specifier.

The selection of the insulation system best suited for a particular building depends on many factors, including all of the items detailed in Figure 4.2. There are other considerations as well. Supply and price are certainly major factors. In fact, cost considerations or availability may play as large a part in the selection of the insulation system as the thermal efficiency, durability, or other attributes.

An important criteria that also applies to all other aspects of building design is: Does the insulation chosen "fit" into the rest of the system? Roof insulation, in order to perform its intended function for the life of the system, must be compatible with all other elements of the roof deck, membrane, and accessories.

CHAPTER FIVE

Flashing, Drainage, and Penetrations

CHAPTER FIVE

FLASHING, DRAINAGE, AND PENETRATIONS

Flashing is one of the most neglected elements of the roofing system in terms of the attention paid to material selection and installation details. Common flashing applications occur at the perimeter edge, curbs, drains and scuppers, equipment supports, vents and flues, and other penetrations. An area of particular concern is the deck and roof termination at a wall or parapet, where proper treatment of flashing can be difficult, but is crucial. One requirement is that the roof should be separated from the structure by kiln-dried wood blocking to prevent deformation and tearing of the base flashing. Figure 5.1 illustrates the basic components and configuration of a base flashing system.

Flashing Criteria and Installation

The roofing specifier should ensure that flashing materials are:
- Impermeable
- Flexible
- Compatible with both the roof membrane and the adjoining surfaces
- Stable
- Durable

Flashing should extend eight inches above the *water line* (defined as the *maximum expected level of water on the rooftop*). The edge should be counterflashed, and the flashing should be fastened every eight inches to prevent slippage (see Figure 5.1). During the installation, the base flashing should be checked by a knowledgeable inspector to ensure proper installation of this most critical element. Flashing is either mopped in place, troweled on with mastic, or fused using heat or solvent.

Base Flashing

Materials
Each type of roofing material requires a flashing material with similar characteristics to assure a good seal. The following is a brief discussion of some of the major flashing materials currently in use.

Asbestos Felts: The Environmental Protection Agency (EPA) has recently begun paying a great deal of attention to the elimination of asbestos fibers from buildings. New restrictions have effectively eliminated the use of asphalt-saturated asbestos felts, which were ideal as base flashing or as support for base flashing in built-up roofs. Being inorganic, the asbestos felts were not susceptible to rot or decay when wet, as are traditional organic felts. When the asbestos felts were used, standard base flashings were protected by a surface ply of granule surfaced roll roofing. The asbestos felts could be molded to conform to the wall and cant configurations, holding their shape well. With asbestos no longer available, new methods must be devised to work with the alternate materials.

Modified Bitumen Felts: As use of asbestos felts declined, modified bitumen-saturated sheets arrived from Europe. These modified bitumens are tough, flexible, and have established a good "track record." The modified sheet consists of a superior plasticized asphalt designed to be torched in place. The sheet is reinforced with a polyester or glass fiber core. This material is compatible with the glass fiber and asphalt systems currently in use for built-up roofing. However, it will take some time to

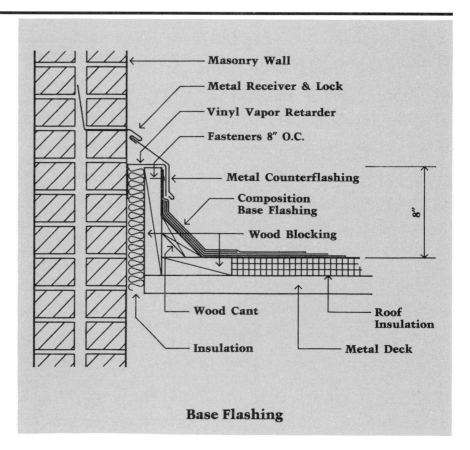

Base Flashing

Figure 5.1

48

educate roofing mechanics in the different torching techniques required for modified bitumen.

Single-Ply Sheet: Each type of single-ply roofing system requires its own compatible flashing material. A typical arrangement is shown in Figure 5.2. However, each manufacturer specifies its own particular flashing methods and details for single-ply systems. These details should be incorporated into the project contract documents as a matter of record.

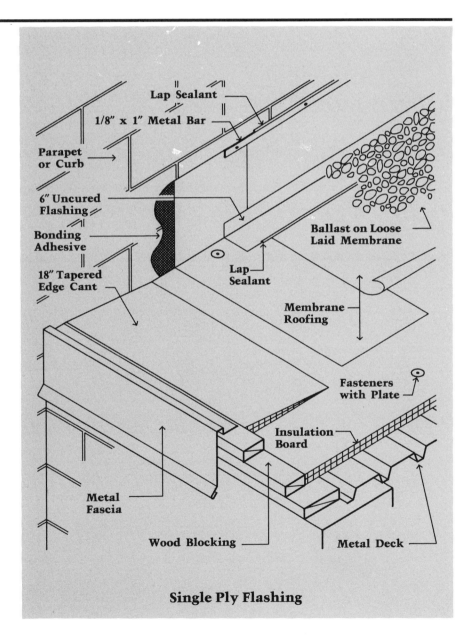

Single Ply Flashing

Figure 5.2

Single-ply technology is in a constant state of flux. It is, therefore, imperative that the designer refer to manufacturer's approved and/ or recommended details for flashings, as of the date of application of that particular system.

Metal: Depending on the roofing system used, metal flashings may be fabricated using aluminum, copper, lead-coated alloy, and galvanized sheet metal. Typically, metal base flashings are used with shingle, slate, and tile roofs, as well as with metal roof systems. Figure 5.3a-c illustrates some of the more common metal flashing details.

Glass Fiber: Glass fiber felt makes very poor flashing material because it has a "memory." Glass fiber is inclined to spring back into its original shape, moving out of position when it is molded around corners.

Cants

Wood fiber cants have proven to be dangerous in that they ignite under the torch or smolder, sometimes causing fires. As a result, perlite has become the recommended material for wall flashing when using torched modified systems.

Flashing at the Perimeter Edge

The proper method of raising the edge flashing out of the plane of the roof is illustrated in Figure 5.4. Using this method prevents water from pooling at the roof edge. Since metal flashings and roofing materials have different coefficients of thermal expansion and contraction, the bond between flashing and roof membrane

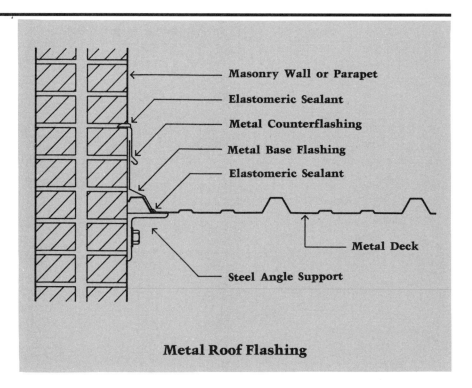

Figure 5.3a

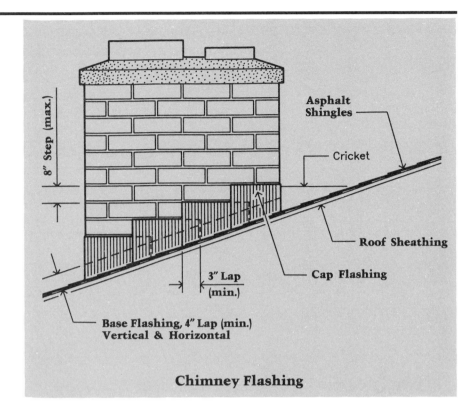

Figure 5.3b

8" Step (max.)

Asphalt
Shingles

Cricket

Roof Sheathing

3" Lap
(min.)

Cap Flashing

Base Flashing, 4" Lap (min.)
Vertical & Horizontal

Chimney Flashing

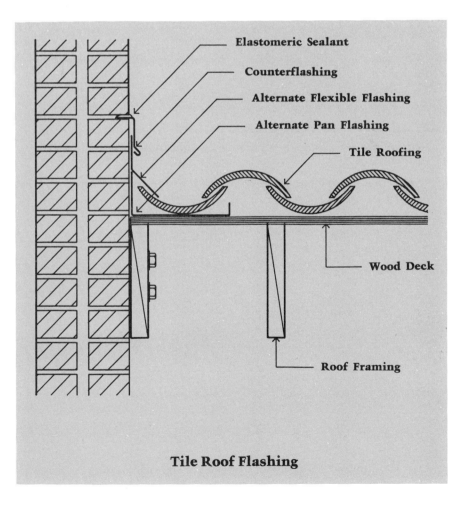

Elastomeric Sealant

Counterflashing

Alternate Flexible Flashing

Alternate Pan Flashing

Tile Roofing

Wood Deck

Roof Framing

Tile Roof Flashing

Figure 5.3c

will invariably break, providing a path for moisture into the system. By raising the metal flanges and providing cover plates at the joints, the resulting damage is minimized. If drainage is required, scuppers, heads, and downspouts should be installed. If possible, *crickets* (flashing systems used to divert water) should be placed between the scuppers to prevent standing water behind the gravel guard (see Figure 5.5).

If metal edging is used, it should be cleated at eight inches on center if there is any possibility that it might be lifted up by wind. For buildings over two stories high and for a fascia of over six inches, this cleat is necessary.

Materials

Aluminum: Aluminum has a relatively high coefficient of expansion and contraction and cannot be soldered. It is, therefore, preferable to use soldered galvanized steel, copper, or stainless steel for perimeter edge flashing. Using gauges heavier than 24 or 26 makes it difficult to form the material and to create joints. The industry has seen a growing use of 24-gauge, pre-painted material for fascias and other exposed architectural metals. Although the paint must be removed from the joints before it can be soldered, the joints can be field-painted.

Rubber Edge: Since metal fascia is usually installed by a sheet metal contractor and not the roofer, manufacturers of single-ply systems are now providing *hard rubber edging* to be used with their systems. Using this material, the roof can be completed *and* flashed by the roofer, leaving no question as to who has final responsibility for the system. This approach also has the advantage

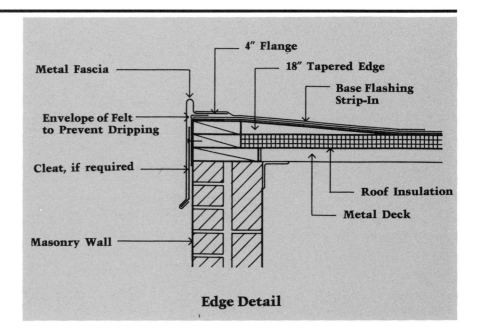

Edge Detail

Figure 5.4

that the entire system is under the roof membrane manufacturer's warranty.

When using rubber edge fascia, it is important that the edging be stripped in with the EPDM flashing. This helps to prevent shrinkage of the material and leaks under the guard.

Drainage Mid-span Placement

Roof drains are too often placed at the highest points on the roof, especially in the case of low-slope and "deadlevel" (flat) roofs. This unfortunate placement occurs because the designer, to save pipe, often places the roof drain directly over a column line, with the drain pipe running down beside the column. As explained in Chapter 2, typical lightweight steel roof structures tend to sag over time, developing ponds between columns. If the roof drains are at the higher points near columns, drainage cannot be achieved. To avoid this problem, an effort must be made to place roof drains in the center of structural bays where deflections are likely to be greatest.

Sumps, Slopes, and Crickets

Drains should always be placed at the deck level, with a 24-inch taper in the insulation to form a sump, as depicted in Figure 5.6. These sumps should be placed in valleys, which can be formed either in the steel or with tapered insulation systems. Tapered crickets may also be installed between drains to ensure positive slopes to drains. Although most manufacturers call for a minimum

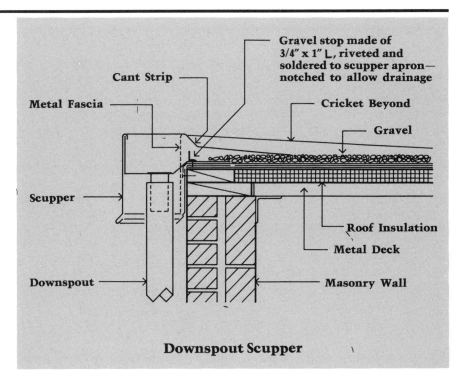

Downspout Scupper

Figure 5.5

$^1/_8$ inch slope, practice has shown that such a slope does not always ensure complete drainage; building construction is not that precise. To be certain of positive slopes, the designer should call for a $^1/_4$ inch per foot taper — either in the structure itself, or by means of tapered insulation.

Drains: Polyvinyl chloride (PVC) drains, shop-made sumps, and cast iron roof drains are all commonly used. PVC drains are compatible with the elastoplastic systems, but are not suitable for torching or hot-mop applications, as they melt under the high temperatures.

Scuppers: When specifying a system, especially ballasted systems, the designer should always include overflow scuppers in adjacent parapet walls to prevent severe damage to the building in the event of a clogged drain.

Maintenance: Regular maintenance is extremely important to an effective, working drainage system. A cost-effective maintenance program may be instituted by the designer, at the end of the construction phase, in the form of strong recommendations for regular checking and cleaning of the roof drains. These measures can be followed by the building manager and can prevent many problems later.

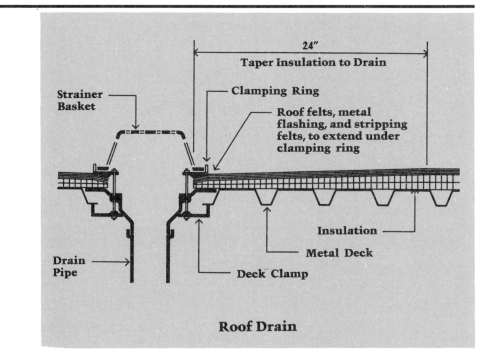

Roof Drain

Figure 5.6

Penetrations

Flues and hot-gas pipes should be flashed in a manner similar to plumbing pipes, using an air space between the pipe and the flashing material to prevent scorching of the flashing membrane. Traditionally, plumbing pipes are flashed using prefabricated metal "jacks" or lead flashing that is turned down into the pipe. Although the alternative, "pitch pockets" are still sometimes shown on drawings, their use should be avoided, as they constitute a continuing maintenance problem. The best way to treat the flashing of small penetrations, such as conduits or lightning arrestors, is by using a hood flashing as shown in Figure 5.7.

Summary

Roof flashing, drainage, and penetrations are crucial elements in a sound roof system. Unfortunately, the importance of these components is often underestimated. The result may be hasty planning and coordination efforts and improper location, or poor choice of materials and/or installation methods. Fortunately, there is an appropriate flashing material for every roofing material. Planning and installation should ensure the integrity of this roofing component.

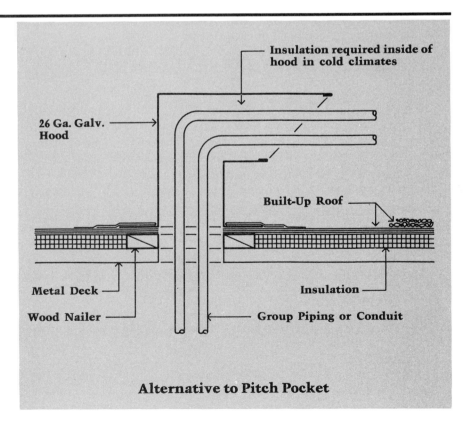

Insulation required inside of hood in cold climates

26 Ga. Galv. Hood

Built-Up Roof

Metal Deck

Wood Nailer

Insulation

Group Piping or Conduit

Alternative to Pitch Pocket

Figure 5.7

CHAPTER SIX

BUILT-UP ROOFS

CHAPTER SIX

BUILT-UP ROOFS

In September, 1987, The National Roofing Contractors Association (NRCA) published a report based on data collected from 131 of its members. Asked what type of roofs they were installing, they replied that 43% of their installations were traditional built-up roofs (BUR), and 57% were single-ply membranes or modified bitumen systems. When asked to list their "problem" jobs, 49 replied "BUR jobs," and 120 noted that the modified bitumen and single-ply systems had developed the most frequent problems. These responses suggest that, contrary to popular belief, the advent of "modifieds" and single-ply systems has not, by any means, solved the industry's roofing difficulties.

As shown in Figure 6.1, built-up, or multi-ply, roofs offer additional layers of protection against moisture penetration. This feature, together with the bitumen's ability to seal itself during warm weather, makes BUR an excellent waterproofing system. Its many years of use, and well-known "track record," have made it the "traditional" roofing material in the U.S. Consequently, roofing journeymen are familiar with the materials and installation methods used for BUR.

In the construction of a BUR, the felts act as reinforcement for the thin layers of bitumen. The bitumen is the actual waterproofing agent. These alternating strata of material form the BUR *membrane*. There are several choices of available felt materials, the most common of which are described in the following paragraphs.

Roofing Felts

Asbestos
Asphalt-saturated 15 lb. asbestos fiber felt is an organic felt that has been used extensively in smooth surface applications. It was once the most popular felt material. However, asbestos felt is no longer used for two reasons: its low tensile strength, and its hazardous effects on human health. In fact, asbestos is being removed from buildings under the direction of the Environmental Protection Agency —either by roofing contractors or by asbestos abatement contractors, at great expense to owners.

Glass Fiber

For a long time, the "workhorse" felt of the industry was 15 lb. organic (rag/paper) saturated felt. Saturated felt is still in use, but the organic felts have been all but replaced by glass fiber felts. Glass fiber felt is available as either *ASTM Type IV* chopped strand saturated felts, or continuous strand, premium grades, with higher tensile strength. In the near future, a *Type VI* felt, with even better performance than premium grade, is to be introduced. However, since splitting is rare in Type IV glass fiber felts, and they work well with the available bitumens, the proprietary Type VI felts may find a limited market.

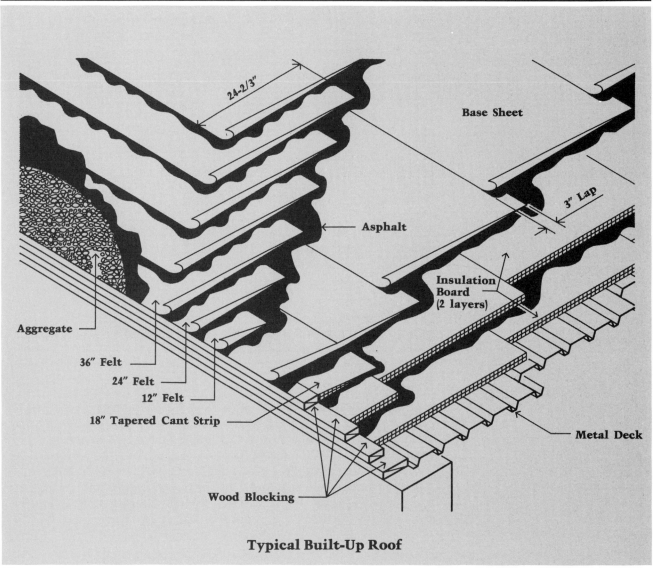

Typical Built-Up Roof

Figure 6.1

The increased use of glass fiber felts (which represent 92% of all felt currently used) has reduced the incidence of blistering, ridging, and splitting. Blistering has diminished because the felts are more porous, allowing moisture to "percolate" up through the felt. Fewer splits occur because of glass fiber's higher strength in both lateral and longitudinal directions. Of course, when improperly installed, channels, voids, and "fishmouths" can still compromise the system. Some bitumens may not be compatible with asphalt-saturated glass fiber felts. For example, there is growing evidence that non-fuming coal tar pitch (Type IV) may be incompatible with glass fiber felts, causing migration of the bitumen from between the felt plies.

Installers should generally stay off a roof until the asphalt has set. This is especially true when glass fiber felts are used. Otherwise, foot or wheel traffic will squeeze the asphalt from between the plies, creating voids.

Base Sheets: Because glass fiber felts are porous and require a smooth, regular surface, a *base sheet* is often used over the substrate. This base sheet is composed of either an organic or glass fiber core saturated and coated with bitumen. It weighs between 35 and 43 pounds per 100 square feet, depending upon the core material used.

Polyester Mat
Another felt material, rarely used in traditional BUR's because of its high cost, is the polyester mat. Its strength is superior to that of glass fiber felts and it is compatible with both hot bitumens and cold adhesives. Polyester mat is used primarily in the manufacture of modified bitumen membranes.

Bitumens In a built-up roof, the actual waterproofing agent is the *bitumen*, a thermoplastic material that "welds" the three or four plies of felt into a membrane. The felts reinforce and stabilize the bitumen layer. The proper temperature is critical to the application of bitumen [See the discussion of equiviscous temperature (EVT) in Chapter 2.] The felts must also be dry and free of wrinkles and imperfections. Foreign matter (such as gravel) must not be allowed to lodge between the felts.

Application Temperatures
Heating bitumen to very high temperatures for long periods of time (over four hours) reduces the asphalt softening point, and raises the coal tar softening point. Concern over these properties has resulted in the practice of using temperatures lower than EVT. The negative effects of these lower temperatures include thick moppings, poor adhesion, splitting, slips, and voids. The opposite problem, excessively high temperatures (above EVT), may lead to incomplete coverage of the film, voids, and the consequential penetration of moisture into the membrane. Figure 6.2 shows the preferred range of application temperatures (EVT) for each type of asphalt and coal tar. In built-up roof installations, the mopping temperatures are critical to the performance of the roof. Therefore, an accurate kettle temperature gauge is an obvious requisite.

Thickness of Bitumen Coatings

If there is good drainage, a thin coat of Type III Steep Asphalt will provide a measure of protection for the roof membrane. For this *flood coat*, the roofing mechanic should remember that *more is not better*. A thick application (over 30 pounds per 100 square feet) is subject to cracking ("alligatoring"). Thick coatings also tend to *slump*, or run down the slopes and into the gutter or over the gravel stop. Most manufacturers' specifications call for a coating of 15 to 20 pounds per 100 square feet, or just enough to solidly coat the surface of the roof. If the pitch of the roof is more than 3:12, *Type IV*, or *Special Steep Asphalt*, should be used (see Figure 6.2). For these steep roofs, the use of asphalt emulsion applied at three gallons per 100 square feet, or fibrated aluminum roof coating applied at two gallons per 100 square feet offers excellent protection. After the surface has weathered, emulsion or fibrated roof coatings are preferred for maintenance or restoration.

Coal Tar Versus Asphalt

The major material choices for BUR surfacing are coal tar and asphalt. A larger quantity is required for coal tar pitch than asphalt, for a given specification. Where 170 pounds of asphalt bitumen meet a certain specification, the same coverage would require a larger quantity of coal tar bitumen. Coal tar bitumen also costs about 50% more than asphalt, per ton. Therefore, there is a significant price difference in the use of coal tar versus asphalt, due

Table of Roofing Bitumen Properties

	Softening Point		Max. Htg. Temp.
	Min.	Max.	
Asphalt ASTM D312-78			
Type I Dead Level Asphalt	135F	151F	475F
Type II Flat Asphalt	158F	176F	500F
Type III Steep Asphalt	185F	205F	525F
Type IV Special Steep	210F	225F	525F
Coal Tar Pitch ASTM 450-78			
Type I Coal Tar Pitch	126F	140F	425F
Type II Waterproofing Pitch	106F	126F	425F
Type III Coal Tar Bitumen	133F	147F	425F

Figure 6.2

to both the quantity required, and the cost of the bitumen. In many areas, the labor cost is also higher for coal tar than for asphalt, due to the noxious fumes it emits.

Surfacing

Most roofing specifications require that a surfacing material be installed after the felts are laid. The purpose of the surfacing material is to protect the felts from direct sunlight, severe weather, fire, and impacts. Roofs without surfacing develop tiny cracks in the asphalt ("alligatoring"), caused by ultraviolet rays. This network of cracks may lead to "ponding" (water collecting in surface depressions) and may eventually destroy the roofing membrane.

Surfacing materials also provide reflectivity, thereby lowering surface temperatures and insuring longer membrane life. Surfacing yields a "thicker" flood coat as the bitumen fills the voids between pieces of aggregate.

Aggregate

Aggregate (gravel) surfacing is the method most commonly used for the protection of membranes. However, gravel surfacing should not be employed on roofs with a pitch greater than 3:12. From the viewpoint of insurance loss prevention, gravel or slag surfacing is much preferred over smooth surfacing because of its fire protection, hail-shielding properties, and low slippage.

Aggregate used for surfacing should be opaque, about 3/8 inch average diameter, and well-embedded in the flood coat. It should meet ASTM Standard D-1863 for aggregate surfacing. Limestone is sometimes used for surfacing, but should be generally avoided as it is relatively soft and tends to dissolve. River bed pea gravel contains shells and mud. This type of material should be washed often and any shells removed during grading.

Some inspectors require that aggregate surfacing immediately follow the installation of the ply sheets. This requirement was valid in the case of organic felt installations, as the sheets curl if left exposed. However, since most felts are now glass fiber, it is advisable to install the plies well ahead of the surfacing operation, so the bitumen may set. This lag time also prevents gravel from entering the space between the plies.

Aggregate is traditionally embedded in a flood coat of bitumen and applied with a dipper or mechanical spreader. The aggregate should be "cast" (spread) into the flood coat either with shovels or mobile spreaders. Unless the spreading is done skillfully, soon after the flood coat, the bitumen can cool and harden, the aggregate insufficiently embedding in the bitumen. Erosion of the aggregate can result, leaving bare bitumen.

The logistics of the aggregate spreading operation are critical. It is important that aggregate be on hand when it is required. Most roofing contractors use conveyors with material hoppers set up beside the building in order to facilitate the movement of material to the point of application on the roof.

One manufacturer of glass fiber felts provides a specification that allows no surfacing. The top ply is left exposed with the idea that inorganic felt will sufficiently resist the weather. This specification invites problems for the following reasons. In practice, it is difficult to prevent some ponding on the roof. Standing water can freeze in winter and enter the porous felts, eventually ripping the roof surface. Fissures soon develop, which promote leaks. Even though water may not "pond" on the rooftop, pinholes and light moppings can cause damage to the interply bitumen beneath the unsurfaced top ply, creating leaks.

Basic Rules of Successful Built-up Roofing

If certain basic principles are upheld, BUR should perform well for 20 years or more. The following guidelines offer a good starting point.

Rule 1: Slope the Roof

The first rule of successful roofing is: never design a flat or "dead level" roof! As is outlined in detail in Chapter 2, the design of near-level or dead-level roofs will eventually result in ponding and deterioration of flashing and roofing materials.

Rule 2: Inspection

The second rule is: provide *continuous* inspection during the placement of roofing, by a knowledgeable inspector. Manufacturers' published specifications have been developed through experience in the field. Carefully following such specifications should lead to a serviceable built-up roof. Proper preparation of contract documents and carefully designed roof details also contribute to the quality of the roof installation, though they cannot ensure it. Quality control must be ensured at the point of "manufacture," *on the roof*. Indeed, there is no substitute for continuous inspection by a *knowledgeable* person, together with a conscientious use of thorough documentation. In *Quality Control In The Application Of Built Up Roofing*, the National Roofing Contractors Association (NRCA) advises, "...a thorough, continuous visual inspection by a knowledgeable person during application is far preferable to test cuts as a means to monitor the roof application."

Rule 3: Restore Temporary Roofs

The third rule is: *always* restore temporary roofing before applying the permanent built-up system. The practice of installing the first plies and one "mopping," then allowing other trades to use it as a staging area, and finally mopping in the additional plies and ballast, has become common as time becomes a crucial factor on construction projects. If a temporary roof is required, it should be properly restored before the permanent roof is installed (see "Temporary Roofs" in Chapter 2).

All accessories, curbs, and penetrations through the roof should also be in place before the roof is installed.

Rule 4: Keep Materials Dry

The fourth rule is: protect roofing and insulation materials from moisture *before*, as well as *during*, construction. The system is susceptible to moisture before, during, and after the roof installation. Prior to installation, roofing materials (especially insulation) must be protected from moisture infiltration. Simply covering these materials may not be sufficient. Some insulations have been found to absorb moisture while stored in a warehouse, to the extent that they are too "wet" to install. It is important to store material off the floor or ground on pallets, and cover it with tarpaulins. The lightweight plastic wrapping that the materials are shipped in is usually not a sufficient covering because it tends to tear easily.

Wet Decks: Lightweight concrete decks are always a concern to the roofing contractor. It is important that they be properly vented. (See "Venting Procedures" in Chapter 2 and Appendix D.) Most specifications call for the roofing contractor to "approve" the deck condition before plans for proceeding with roof installation. The roof installer should perform a deck dryness test. (See Appendix C.)

Rule 5: Avoid the Void

The fifth rule of successful roofing involves the prevention of voids between plies in a BUR. The most prominent cause of failure in BUR's according to an NRCA survey was interply blistering.

To prevent voids, an even bitumen application is essential; this can be obtained when the application temperature of the bitumen is in the equiviscous temperature range (EVT). The EVT is usually printed on the asphalt container. The roofing contractor must heat the bitumen in the kettle to a temperature above the EVT, but below the flash point, in order to allow for the chilling of the asphalt as it is transported and applied. The bitumen should "roll" (spread out) along the laps on application, and the plies should be installed "shingle fashion" as shown in Figure 6.1.

Summary Despite the popularity of single-ply roofing systems, built-up roofing continues to be a mainstay in contemporary construction. BUR offers the specific advantages of additional layers of protection from moisture penetration, and an ability to seal itself in warm weather. Since it has such a long history, it has a proven "track record". Over the years, roofing installers have become familiar with this system and BUR manufacturers have had the opportunity to improve upon its basic materials.

CHAPTER SEVEN

MODIFIED BITUMENS

CHAPTER SEVEN

MODIFIED BITUMENS

The *modifieds* were developed in Europe during the sixties and introduced in the U.S. in the early seventies. The term *modified* refers to the addition of plastic or rubber-based polymeric binders to asphalt to improve its performance and weatherability. *Modified bitumen* is used in multiple layers in what is essentially a "factory assembled" built-up roof. As shown in Figure 7.1, several coats or laminations of modified bitumen are reinforced with a

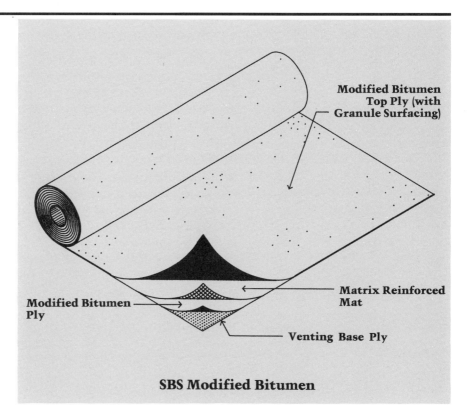

Modified Bitumen
Top Ply (with
Granule Surfacing)

Matrix Reinforced
Mat

Modified Bitumen
Ply

Venting Base Ply

SBS Modified Bitumen

Figure 7.1

woven glass or polyester fabric reinforcing mat, which is often covered with a granule-surface *cap sheet*. Modified bitumen roofing exhibits many characteristics of a built-up roof (BUR), yet affords the ease of application of a single-ply system.

Polymer Modifieds

The three most popular bitumen modifying polymers are: SBS (styrene butadiene styrene copolymer), APP (atactic polyproplyene), and SBR (styrene butadiene rubber). Modifieds represent a major area of product growth for roofing materials, with a 25% annual growth in sales volume. By 1987, modifieds had captured over 15% of the low-slope roofing market.

Problems with Modifieds

Some problems with modifieds are now being recognized by regulatory agencies such as the National Bureau of Standards (NBS), and the National Roofing Contractors Association (NRCA). The following list includes some of the major pitfalls that are possible when using this system.

- Although rare, blistering may occur between the base ply and the modified sheet if moisture is trapped or voids are left between the plies. Blistering usually occurs when the asphalt between the plies is mopped at a low temperature (significantly less than 425 degrees Fahrenheit) and there is an inadequate seal.
- Slippage is possible, particularly on steep slopes when using asphalt at a low melting point.
- Membrane flashing may separate if cant strips are eliminated. The flashings are usually torched on, and the fear of fires has led some designers to omit cants. Fires can be avoided in this process if perlite or incombustible cants are specified in lieu of wood cants.
- The most common problem with SBS modified systems is the delamination of the modified membrane due to improper temperature of the asphalt at application.

SBS Modified Membranes

A trend is currently underway for the increased use of SBS modified membranes. SBS modified membrane is generally mopped down over a base sheet in a hot mopping of conventional, blown asphalt. This system offers the advantage of similarity to traditional BUR systems, and therefore requires little retraining of installers.

Modified bitumens are not, strictly speaking, single plies. They are "hybrids"—neither single-ply, nor BUR, but exhibiting the advantages of both. Modified bitumens are applied like a traditional BUR, but with only one "ply,"thereby reducing labor cost. The modified bitumen's reinforced construction gives it a plasticity similar to that of single-ply systems with the "self-healing" and multi-ply safety characteristics of BUR.

A single "ply"of modified bitumen membrane is not sufficient to ensure a complete roof system. A base sheet should first be applied to the substrate, to provide a uniform substrate surface and serve as a back-up water stop. Venting base sheets are recommended for use in conjunction with roof venting systems; vapor retarder base sheets are also available. These sheets act as a substrate for the modified sheet, and serve a secondary role as a vapor control or a means of venting. Although these membranes have excellent weatherability and good resistance to ultraviolet radiation, a protective surface is also recommended. Most modified bitumen material is available with a factory-applied mineral "cap sheet" or aggregate topping.

Torched APP Membrane

Torched Atactic Polypropylene (APP) membrane is installed using a series of propane torches to evenly melt the underside of the membrane. This multi-torch unit includes "soft" wheels and an axle that allows the membrane to unroll as it is being heated by the torches. This process reduces labor costs and provides better control of the application temperature. On cold days, it is critical that the membrane be stored in a heated space prior to installation so that it is pliable enough to roll out evenly on the roof.

The APP system offers the following advantage: It is not hot-mopped and therefore requires no kettle. Consequently, it is ideal for roofs with access problems, such as high-rise buildings, or for jobs with numerous small roof areas or levels.

The use of APP modified does present certain potential pitfalls. For example, rupturing and splitting caused by the use of weak reinforcements have been reported as a problem with APP modifieds. There is also a possibility of fires resulting from combustible substrates which may smolder while the roofers are present, and ignite after they have left. Furthermore, the use of torches requires skill if bitumen temperatures are to be properly regulated.

SBR Modified Membranes

The first SBR introduced in the U.S. was reinforced with a polyethylene mat and had an aluminum foil facer. This material was used widely in the 1970's but, because of its inferior elongation characteristics and fatigue endurance, it has been eclipsed by APP, and especially by SBS modified bitumen.

Coatings

Surfacing is intended to protect the membrane — not only from the weather, but also from roof traffic and other forms of abuse. Most manufacturers call for a protective coating or include in the manufacturing process a surfacing for their membranes. Although modified membranes offer good resistance to ultraviolet radiation degradation, they must still be protected. Some APP manufacturers still offer "unprotected" membranes, but the industry is shifting toward the requirement of some sort of coating over all systems, whether manufactured or job-applied. These can be traditional hot asphalt and aggregate, cold asphalt emulsions, asphalt-fibrated coatings, or granules in adhesive.

Application Recommendations

It may be helpful for the specifier to incorporate into the specifications a copy of the document, *Quality Control Recommendations for Polymer Modified Bitumen Roofing*, jointly published in 1988 by the National Roofing Contractors Association (NRCA) and Asphalt Roofing Manufacturers Association (ARMA). Some major points of this document are reproduced in Figure 7.2 for reference by owners, specifiers, and installers.

Summary Modified bitumens offer the known characteristics of asphalt bitumens, with the added benefits of better performance and resistance to weather. Of the three basic types of modifieds, SBS and Torched APP membranes are the most practical and advantageous. Both have potential pitfalls, which can, for the most part, be avoided by careful adherence to manufacturers' recommendations for installation.

Quality Control Recommendations
for Polymer Modified Bitumen Roofing

A. Introduction
This document covers:

- Application recommendations (workmanship)
- Examination of modified bitumen roofing during construction

This document is not intended for examination or evaluation of *existing* roof systems.

Roofing with modified bituminous membrane materials is a construction process. In any construction process, it is difficult to place precise limits on the quantities and dimensions of materials applied and to evaluate conformance to those limits. Roofing with modified bituminous materials is no exception.

The construction of a successful modified bituminous roof involves the skillful and systematic arrangement of many components under varying temperatures, weather and job site conditions. The variety of material combinations and different roof penetration and perimeter details make it difficult to construct and analyze a roof system assembly. These facts must be considered when the roofing process is examined and evaluated.

A successfully performing roof assembly is the result of a collective commitment by those involved. These parties and their responsibilities include:

1. The owner: for a realistic financial commitment to obtaining a good roof and for regular maintenance after construction.
2. The designer and specifier: for the overall roof system and building design, which take into account the effect of all building components on roof system performance.
3. The material supplier: for development of sound modified bituminous specifications and for the manufacture of roofing materials that meet current industry standards for such materials.
4. The general contractor: for proper coordination of construction and supervision of other trades' work that affects roof performance.
5. The roofing contractor: for proper application in accordance with the applicable specifications and for adherence to good workmanship practices.

Among the most critical factors in roofing performance are:

- Building design
- The substrate for the roofing membrane
- Specifications
- Physical properties of the materials
- Proper storage and handling of materials

Designers should pay particular attention to slope for drainage, flashing details, expansion joints (as appropriate) and conditions to which the roof may be subject after installation. All have a substantial impact on roof performance.

The recommendations that follow are based on a careful study of applicable industry literature and the collective experience of roofing contractors and roofing material suppliers.

B. Examination of Application
The most effective way to evaluate the quality of a modified bitumen membrane application is by thorough, continuous visual examination at the time of application, conducted by a person with a knowledge of roofing technology and good workmanship practices.

The term "visual examination", as used throughout this document, is not intended to preclude the use of such obvious tools as measuring tapes and thermometers. These are tools that can aid in the visual examination and their use is encouraged. A person deemed knowledgeable is any individual who because of application experience and general roofing industry knowledge is competent to conduct the procedures described in this document. This individual might be a roofing superintendent or foreman, a contractor, a designer, or an owner's or material supplier's representative.

The recommendations presented are intended to aid in the examination of modified bituminous roof membrane systems as they are being applied. The individual responsible for visual examination during application should check to see that:

1. The roofing system materials are properly stored and are in a condition satisfactory for application. Materials should be examined for defects. Roll goods should be stored off the ground, standing on end and protected from the elements. Material that is damaged or wet should not be used.
2. The surface of the deck is clean, firm, smooth and sufficiently dry to allow for proper roof application.
3. Edge nailers, curbs and penetrations (including drain bases) are in place and properly secured prior to roofing so that the roof system can be installed as continuously as possible.
4. The weather and job conditions are suitable for roofing.
5. There is positive attachment of the insulation or base ply to the roof deck.
6. The bonding agent used for the assembly of laps and for adhering the modified bituminous membrane to the roof substrate is permitted and acceptable to the material supplier.
7. Fully adhered materials are embedded into a continuous firmly bonding film of molten modified bitumen compound, asphalt or adhesive. The bonding agent should flow out at the seams.
8. Positive attachment is achieved with partially adhered and mechanically fastened modified bituminous systems, which preclude a continuous film.
9. The roof membrane is applied in order to allow water to run over or with side and/or endlaps in the membrane.
10. Sheets are aligned so that minimum required end and side lap widths are maintained.
11. Membrane laps are fabricated watertight. Voids and fishmouths within the lap should be repaired as soon as possible.
12. Temporary water cutoffs are installed at the end of each day's work and are removed prior to installing additional insulation or membrane.

Figure 7.2

13. Surfacings, when required, are applied in accordance with the membrane material supplier's requirements and are applied so that the entire membrane surface is covered.

C. Substrate Application Recommendation

There are a variety of substrates to which the modified bituminous membrane can be applied. These include existing roof membranes, insulations and decking materials. Each of these is unique, and has its own application guidelines and precautions. The condition of the substrate is critical to the performance of the membrane.

1. Decking Materials

The surface of the deck shall be clean, firm, smooth, and visibly and sufficiently dry for proper roof application. There are different preparation requirements for new construction, replacement or re-covering over each specific type of substrate. The roofing material supplier and the roofing contractor can be responsible only for acceptance of the surface of the substrate that is to receive the roofing system. Design considerations and questions of structural soundness must be determined by the owner or the owner's representative.

The important application factors are:
- If there is a question about the suitability of the deck and/or its surface, the material supplier, the roofing contractor and the owner must all reach agreement before work is to proceed.
- When primer is required, the surface shall be primed with an amount sufficient to cover the entire surface. The amount required will vary, depending on the nature of the surface. The primer should be allowed to dry before membrane application.
- If visual examination indicates deficiencies in primer coverage, additional primer shall be applied for complete coverage, and allowed to dry before membrane application.

2. Existing Roofs

The surface of an existing roof should be suitable to receive a modified bituminous roofing membrane.

NRCA and ARMA recommend the use of a re-cover board over existing gravel-surfaced roofs before the application of the membrane system. Loose and protruding aggregate can be removed by spudding, vacuuming or power brooming to achieve a firm, even surface to receive the re-cover board.

When the existing roof is smooth or mineral-surfaced, refer to the material supplier's specifications. Material suppliers' specifications vary; some permit fully adhering by heat welding, hot mopping or adhesives; others require base sheets over the existing roof; while others permit spot mopping, strip attachment and/or mechanical fastening.

When the use of primer over the existing roof is required, refer to the primer recommendations covered in Section C, Subsection 1, Decking Materials.

3. Preformed Roof Insulation

Manufacturing tolerances, dimensional stability, variables during application, and the nature of insulation boards make it difficult to obtain tightly butted joints, and some variance is expected.

The important factors are:
- The spacing between each insulation board should range from no measurable space (firmly butted) to a maximum space of 1/4" or as specified by the material supplier.
- Insulation gaps between adhered or mechanically fastened insulation boards in excess of 1/4" should be filled with roof insulation. Gaps between loose-laid insulation boards should be reduced by adjusting the boards or by adding insulation.
- If the insulation boards appear to be out of square, a diagonal measurement shall be made to confirm the squareness. Defective material should not be used.

4. Base Plies

The application of the base ply shall result in a clean, firm and smooth surface to allow for proper membrane application.

The base plies specified by the modified bitumen material supplier can be installed using mechanical fasteners (where appropriate), hot asphalt or adhesives (fully or partially adhered).

The important application factors are:
- The application should result in a firmly bonded membrane.
- Base plies should be aligned so that minimum specified end and side lap widths are maintained.
- Products are frequently supplied with laying lines. These should be used to position the subsequent course of material for compliance with nominal lap requirements. Because of the variation in substrates, products and systems, the material supplier's specification shall be referenced for minimum requirements.
- Should misalignment of the sheet occur during application, causing a variation from the minimum lap requirements, the sheet should be cut and realigned to satisfy the minimum lap requirements.
- For those products without laying lines, the material supplier's stated lap dimension shall be the minimum lap requirement.

5. Fastener Spacing

Various mechanical fasteners are used in the application of insulation and base plies, and may also be used to attach the modified bituminous membrane to the roof deck and to attach flashing materials. They are most often specified by type, number and spacing distances to fulfill these attachment functions. Practical considerations often prevent exact spacing (e.g., 6 inches on center). Reasonable variances from specified spacing distances should be expected.

The important application factors are:
- The minimum number of fasteners should be as specified.
- The fastener type and spacing should be as specified by the material supplier with the understanding that the spacings are average values and the spacing between any two fasteners may vary.
- Should fastener deficiencies be discovered, additional fasteners shall be installed as needed and spaced appropriately.

Figure 7.2 (cont.)

D. Membrane Application Recommendations

Modified bituminous roof membranes are applied using a variety of methods, including heat welding, hot asphalt and adhesive applications. Sometimes more than one method of attachment is employed in a single system, such as at flashing areas. The degree of heat required for appropriate heat welding, the variables affecting the application rates and temperatures of any asphalt used, or the amount of adhesive used, are subject to many variables, including weather conditions, job conditions, material type and application method. In addition, the material supplier's requirements must be considered.

1. Laps and Seams in Membrane Construction

The strength and waterproofing characteristics of modified bituminous roof membrane systems depend on the construction of laps and seams between membrane sheets and between membrane and flashing materials. Lap dimensions are typically specified as an amount of overlap between adjacent sheets. Because of the many construction variables already mentioned, some variance may occur. An overlap exceeding the specification is not considered to be detrimental. However, the material supplier's minimum lap values must be maintained so as not to affect the strength and waterproof integrity of the membrane.

The important application factors are:

- Membrane laps should be fabricated watertight. Voids and fishmouths within the lap should be repaired.
- Products are frequently supplied with laying lines. These shall be used to position the subsequent course of material for compliance with nominal lap requirements.
- Because of the variation in substrates, products and systems, the material supplier's specification shall be referenced for minimum requirements.
- For those products without laying lines, the material supplier's stated lap dimension shall be the minimum lap requirement.
- If examination reveals insufficient lap width, a strip no narrower than twice that of the required lap shall be installed over the deficient lap, using the appropriate attachment method.
- All laps of heat welded systems should be checked for adequate bonding. Any unbonded area should be sealed.

2. Fully Adhered Membrane Application

a. Heat Welded Applications

Heat welding equipment is used with many modified bituminous membranes to heat the underside of the sheet, melting enough of the modified bituminous compound to achieve adherence to the underlying substrate. The heat welding application shall result in a flow of bituminous compound from the membrane lap to form a seal.

The important application factors are:

- Sufficient heat should be used to bring the modified bituminous compound to a molten state to ensure that the bituminous compound will form a continuous, firmly bonding film for fully adhered systems.
- Areas of loose membrane should be cut, reheated and bonded to the substrate. An additional piece of the same material used in the original membrane shall be installed beyond the area to be repaired, extending a minimum of 6 inches over the cut in all directions. Some material suppliers may accept alternate repair techniques.

b. Hot Mopped Asphalt Applications

Hot asphalt is used with some modified bitumen membranes to adhere the sheet to the substrate. The asphalt shall be the type specified by the material supplier. During the application of hot asphalt the presence of a continuous, firmly bonding film of asphalt should be observed. Generally, the asphalt should flow out at the seams.

Note: If aesthetic appearance is a factor or dictated by specification, matching granules may be used to cover the flow-out area.

The important application factors are:

- Asphalt shall be applied at a temperature within ± 25 degrees of its equiviscous temperature, unless the material supplier requires other specific application temperatures. Equiviscous temperature is the temperature at which a viscosity of 125 centistokes is attained and is considered the optimum application temperature for hot asphalt applied roofing.
- Areas of loose membrane shall be cut out and removed, or mopped under and replaced into their original positions when suitable. An additional piece of the same material used in the original membrane shall be installed over the area to be repaired, extending at least 6 inches beyond the cut in all directions.

c. Adhesive Applications

Cold adhesives are used with some modified bitumen membranes to bond the sheet to the substrate. The adhesive shall be the type specified by the material supplier. During the application of a fully adhered system the presence of a continuous, firmly bonding film of adhesive should be observed. The adhesive should flow out from the membrane to form a seal.

Note: If aesthetic appearance is a factor or dictated by specification, matching granules may be used to cover the flow-out area.

Areas of loose membrane shall be cut out and removed, or mopped under and replaced into their original positions when suitable. An additional piece of the same material used in the original membrane shall be installed over the area to be repaired, extending at least 6 inches beyond the cut in all directions.

3. Partially Adhered Membrane Applications

In partially adhered systems, modified bituminous compound, asphalt and adhesives are used to secure the roof assembly to the structure. However, these materials are NOT applied in a continuous film. Rather, they are installed in a configuration prescribed by the material supplier (i.e., spot or strip bonding). They are most often specified by amount and spacing distances. Practical considerations often prevent applying the exact amount and spacing (e.g., 12 inches in diameter and 24 inches on center) of the bonding agent and reasonable variances should be expected.

Figure 7.2 (cont.)

The important application factors are:
- The application should result in a firmly bonded membrane.
- The spacing of the bonding agent should be as specified by the material supplier with the understanding that the spacings are average values and the spacing between any two adhesive locations may vary.
- Deficiencies shall be corrected by installing additional bonding agent as needed, spaced appropriately. The scope of the discrepancy must be determined and appropriate remedial action taken.

4. Surfacing Materials

Various materials are used to surface some modified bituminous roof membranes. In addition to their primary functions of protecting the membrane from the elements, these surfacings often serve to increase the fire- and impact-resistance of the roofing system. Liquid-applied surfacing materials vary widely in physical properties and in formulation. Surfacing materials and aggregate are applied by various techniques, such as hand spreading and mechanical application. They are also applied on a variety of roofs in many climatic conditions. These factors and others preclude a high degree of uniformity in applying liquid-applied surfacing materials and aggregate over a roof area.

The important application factors are:
- The surfacing shall completely cover the modified bitumen roofing membrane.
- If visual observation indicates deficiencies in surfacing material and aggregate, apply additional surfacing material and aggregate for complete coverage.

E. Flashing Application Recommendations

Modified bituminous flashing materials are applied using a variety of methods, including heat welding, hot asphalt and adhesive applications. The degree of heat required for appropriate heat welding, the variables affecting the application rates and temperatures of any asphalt used, or the amount of adhesive used, are subject to many variables, including weather conditions, job conditions, material type and application method. in addition, the material supplier's requirements must be considered.

All of the items covered under Section D, Membrane Application Recommendations, are applicable to flashing applications. Refer to Subsections 1 through 4 for specific recommendations relating to laps and seams, heat welding, hot mopping, adhesive applications and surfacings.

1. Substrates

The surface that will receive the flashing shall be clean, firm, smooth, visibly and sufficiently dry for proper application. There are different preparation requirements for new construction, replacement or re-covering over each specific type of substrate.

The important application factors are:
- The surface that will receive the flashing should be clean, firm, smooth, and visibly and sufficiently dry for proper application. If there is a question on the surface's suitability, the roofing contractor, material supplier and owner, or owner's representative, shall reach agreement before work proceeds.
- Some substrates require priming. When priming is required, the surface shall be primed with an amount sufficient to cover the entire surface to receive the flashing. The amount required will vary, depending on the nature of the surface. The primer should be allowed to dry before flashing application begins.
- If visual examination indicates deficiencies in primer application coverage, apply additional primer for complete coverage.
- It is recommended that wood products be covered with a mechanically fastened base ply prior to application of modified bituminous flashing.

2. Attachment

There are a variety of suitable methods of attachment for modified bituminous flashing materials. The recommended procedures (described in Section D, Membrane Application Recommendations) for each type shall be followed to obtain a continuous and firm bond. The material shall also be lapped, coated and counterflashed as prescribed in the material supplier's specifications.

Mechanical fasteners are usually required to prevent flashing materials from sliding. Refer to Section C, Subsection 5 of this document for recommendations.

3. General

The flashing terminations should be located a minimum of 8 inches above the level on the roof; counterflashing should be as specified by the material supplier. Some material suppliers have a maximum height limitation for flashings; refer to the material supplier's recommendations.

A cant strip may be installed to modify the angle between the roof deck and the vertical surface. For heat-welded applications, cant strips should be made from a flame-resistant material and covered with a base felt. The material supplier should be consulted for recommendations regarding the use of cant strips.

The membrane should be installed before the flashing is applied and should be installed above the plane of the roof, above the cant and up the vertical surface as specified by the material supplier. The roof membrane should not be carried up the vertical surface to act solely as a base flashing.

Areas of loose, inadequately or improperly bonded flashings should be removed and reinstalled, using the same type of materials used in the original application.

Laps that are inadequately bonded should be cut out, rebonded and reinstalled.

Figure 7.2 (cont.)

CHAPTER EIGHT

SINGLE-PLY SYSTEMS

CHAPTER EIGHT

SINGLE-PLY SYSTEMS

When the cost of bituminous roofing installations began rising, owners and specifiers sought a more reliable alternative to built-up roofing (BUR). Roofing contractors traveling in Europe returned to the U.S. with stories of new roof systems that used synthetics in a single-ply sheet. The idea of a loose laid "rubber roof" seemed ludicrous. Nevertheless, these single-ply membranes were already in use as pool and pond liners, and as waterproofing membranes, and the European manufacturers convinced roofers to try the new system. By 1980, single-ply roofing (SPR) had captured about 8% of the U.S. roofing market and its acceptance was accelerating. In 1987, the total volume of SPR shipped had reached one billion square feet, about one third of the low-slope roofing market. Figure 8.1 shows a single-ply EPDM roofing system with flashing details.

In Europe, EPDM (ethylene propylene diene monomer) acquired a poor performance history as a roofing material. In the United States, however, the research and development effort expended on single-ply roofing was extensive. Large strides were made in both the material quality and attachment methods. One of the first major EPDM applications in the U.S. was at Chicago's O'Hare airport. Despite the intensive research and testing conducted on SPR systems, the major manufacturers of BUR failed to recognize them as serious competition in the roofing market. Many who held out against SPR saw their profits eroding, and lost their claim to the roofing market altogether. Today, there are over 450 listings of SPR systems in NRCA's *Materials Guide*, and SPR has over 50% of the low-slope market.

SPR's are isotropic (have equal strength in both directions). They offer better moisture resistance than the BUR's (unless torn or punctured), and exhibit superior elasticity at sub-zero temperatures. Compared to BUR, they are also subject to fewer application restrictions during inclement weather. When leaks do occur, they are more easily repaired than is the case with BUR systems. The success of SPR systems has been achieved through extensive manufacturer effort in research, testing, field support, warranty programs, and incentives for quality installations.

Advantages and Drawbacks

Types of SPR

Single-ply membranes do not involve the hot bitumen mixes and mopping procedures that are the trademark of built-up roofs. While their installation is thereby a cleaner process than that of BUR, some systems do involve cements, solvents, or sealants.

The *Roofing Materials Guide*, published by NRCA, lists four major types of single-ply material. One of these, Modified Bitumen Membranes, has already been discussed in Chapter 7. Modified membranes are not true SPR's, but a hybrid. The other dominant SPR types are:

Thermosetting (vulcanized or "cured" elastomers)
- EPDM (ethylene propylene diene monomer)
- Neoprene (chloroprene rubber)

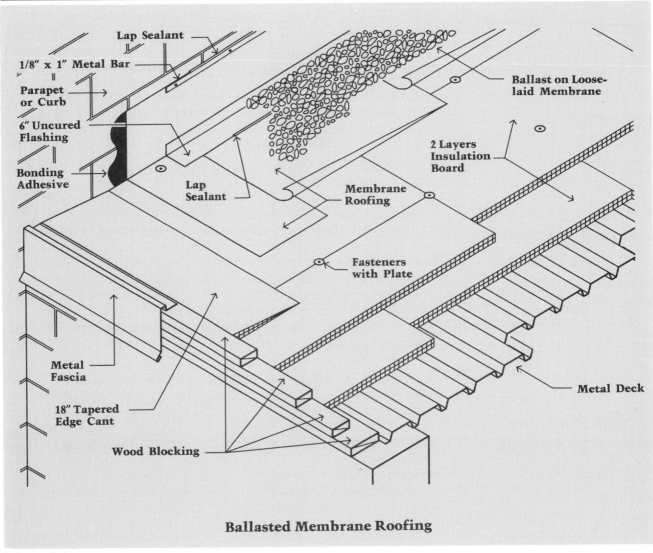

Ballasted Membrane Roofing

Figure 8.1

Thermoplastic (non-vulcanized or "uncured" elastomers)
- CPA (copolymer alloy)
- CPE (chlorinated polyethylene)
- EIP (ethylene interpolymer alloy)
- NBP (acrylonitrile butadiene copolymer)
- PIB (polyisobutylene)
- PVC (polyvinyl chloride)

CSPE (chlorosulfonated polyethylene)
- This is neither a thermosetting nor thermoplastic material.

Most SPR has excellent resistance to heat, ozone, and ultraviolet light, and retains its flexibility at low temperatures. Each type of SPR has some problems with regard to chemical incompatibilities or installation difficulty; none is perfect in all areas. For a given installation, one may refer to Figure 8.2 and select the most appropriate type of SPR. The amount of expected roof traffic, whether or not grease or oil may be present (from, say, restaurant roof exhausters), what size roof area is involved, and whether or not color is important — all are factors to consider.

The most widely used type of single-ply roofing is EPDM. According to the Rubber Manufacturers Association, over one billion square feet of EPDM was sold in 1987. Available in rolls of up to 50 feet wide and 200 feet long, EPDM requires fewer field seams, which are the weakest parts of the system. These large sheets can only be used in a ballasted system, since mechanical fasteners and adhesives are normally placed at seams. Although the ballasted system adds ten pounds per square foot of dead load to a structure, its benefits exceed the extra structural costs.

As demand for SPR systems increased, new manufacturers entered the market. Some of them had only a remote connection to the roofing and waterproofing industry. There are still many in the market trying to compete with the large (mainly tire) manufacturers who formulate and produce their own EPDM membranes.

Selection Some basic guidelines should govern the selection of a proposed SPR system. One such set of guidelines published in the May, 1987 *Engineer's Digest*, is as follows.

1. Does the material delivered to the job site bear the UL label?
2. Does the system pass the Factory Mutual I-90 Wind Uplift Test?
3. How much does the completed system add to the dead load weight of the roof structure — and can that extra weight be tolerated?
4. Does the manufacturer actually manufacture the membrane, or does this company simply market the product?
5. How long has this manufacturer been in the business and what is its track record?
6. What is the net worth of the manufacturer?
7. Can the manufacturer provide a list of installations and references?
8. What is the financial strength of the roofing contractor?
9. Is the roofing contractor a certified applicator (in good standing) who has been trained to install the system?

Single-Ply Membrane Chart[1]

Manufacturer	Brand Name	Product Type*	Oldest U.S. Installation (yrs.)	Color***	Thickness (mils)	Weight (lb./sq.ft.)	Roll width range	Roll length range	Reinforcement	Tensile strength (psi) (ASTM D472 & D751)	Elongation (%) (ASTM D472 & D751)	Low temp. flexibility (%F) (ASTM D746)	Dimensional Stability (%)
Carlisle SynTec Systems	Sure-Seal	EPDM	26	w/b	45	.28	54"/50'	50'/200'	n	1640	500	-75	<2%
Cooley Roofing Ssytems	Cool Top	CPE	11	w	40	.28	72"	108'	y	135	29	-40	1%
Duro-Last Roofing	Duro-Last	Reinf. CP	9	w/t	35	.26	†	†	y	†	27	-40	1%
Firestone Bldg. Products	Rubber Gard	EPDM	8	b	45	.28	60"/50'	100'/200'	n	1305	300	-49	<2%
Goodyear Tire & Rubber	Versigard	EPDM	23	w/b	45	.28	54"/50'	to 150'	n	1300	350	-75	<2%
Gen Corp Polymer Prod.	Gen Flex	EPDM	7	w/b	45	.26	42"/50'	100	n	1500	425	-67	<2%
Kelly Energy Systems	Premium Whaleskin	EPDM	22	w/b	45	.28	10'/20'	50'/100'	n	1400	300	-75	<1%
Kelly Energy Systems	Kelly CPA	CPA	8	w	50	.32	72"	75'	y	250	25	-40	<.5%
Manville Roofing Systems	SPM Roof Systems	EPDM	9	b	45	.26	to 50'	150'	n	1305	300	-49	<2%
Republic Powdered Metals	Geoflex	PIB	10	w/b	100	.57	42"	60'	y	550	410	-40	<1%
Sarnafil	Sarnafil	PVC	12	C	48	.33	78"	65'	y	230	20	-40	≪1%
Seal-Dry	System 7000	CPA	5	W	40	.25	to 20'	100'	y	400	35	-30	≪1%
Seaman	Fiber Tite	EIP	20	t	33	.25	58"/20'	to 100'	y	8500	30	-30	<1%
J.P. Stevens & Co.	Hi-Tuff	CSPE	12	w	45	.29	64"	80'	y	1400	400	-40	≪1%
Trocal Roofing Systems	Trocal Membrane	PVC	15	w	60	.49	†	†	y	2400	250	-40	<1.5%

*EPDM: ethylene propylene diene monomer; CPE: chlorinated polyethylene; CSPE: chlorosulfonated polyethylene; CP (CPA): Copolymer alloy; PIB: polysobutylene; PVC: polyvinyl chloride; EIP: ethylene interpolymer alloy

**Substrates & insulation: 1) perlite board; 2) urethane; 3)phenolic; 4)wood fiber; 5)fiberglass; 6) polystyrene; 7) isocyanurate; 8) cellular glass; 9) existing asphaltic roofing

Figure 8.2

Contaminants to avoid***	Incompatible with (substrate)**	Installation Method				Lap Joining Method						Warranty & Remarks†
		Loose laid/ballasted	Mechanically attached	Fully adhered	Protected membrane	Adhesive/Sealant	Splice tape	Hot air weld	Solvent weld	Pre-fab. self-seal	Factory dielectrics	
A_BC		•	•	•	•	•	•					5,10,15 yr. warranty avail.; UL Class A&B; FM I-60 & I-90
D_F	$^14_8$	•	•					•				10 yr. warr.; UL Class A; FM I-60 & I-90
D		•	•					•				†prefab. up to 2500 sq. ft. sections; ASTM D882:7200 psi; 15,20 yr. warr.; UL Class A; FM I-90
D_E		•	•	•	•	•	•					5, 10, 15, 20 yr. warr.; UL Class A; FM I-60, I-90
E	3_7	•	•	•	•	•						5, 10, 15, 20 yr. warr.; UL Class A; FM I-90
D_E		•	•	•	•		•					5, 10, 15 yr. warr.; UL Class A; FM I-90
E		•	•	•	•	•	•					6, 11, 15, 21 yr. mat'l.; 6, 11 yr. lab/mat.; UL Class A; FM Class 1
$^{G}H_{E_B}$			•					•	•			15 yr. mat'l.; 5, 10 yr. labor; UL Class A&B; FM I-90
$^{B}C_G$	35_8	•	•	•	•	•						5, 10, 15 yr. warr.; UL Class A,B; FM I-60, I-90
$^{B}C_G$		•	•		•					•		10 yr. warr.; UL Class A; FM I-90
$^{B}G_H$	$^134_{5_8}$	•	•	•	•			•				10 yr. warr.; UL Class A; FM I-90
†	8		•					•			•	†contact factory; 15 yr. warr.; UL(A); FM I-60, I-90
†		•	•		•			•			•	†contact factory; 5, 10 yr. warr.; UL(A); FM I-90
C_D		•	•	•	•			•				5, 10, 15 yr. warr.; UL Class A; FM I-90
(1)	3_4	•	•		•			•	•			†varies with membrane type; 10 yr. warr.; (1) contact manufacturer; UL Class A; FMI-90

***Contaminants: A) acids; B) oils; C) solvents; D) hydrocarbons; E) petroleum-based substances; F) oxidizers; G) coal tar; H) asphalt

****Colors: b: black; w:white; t:tan; g:gray; c:colors

Notes: This chart is not intended to endorse any product or manufacturer, but merely to illustrate the range of materials available and their various properties. It is not in any way a complete listing of products available, or intended to replace thorough research by the specifier of available systems and materials. Due to the rate of product development and improvement in the single-ply market, entries may be superseded.

10. Are there enough manufacturer-certified applicators of this material in your area to provide you with competitive bids?
11. Does the system manufacturer provide the warranty?
12. Who performs warranty service?
13. Does the warranty cover the entire system?
14. Is the warranty void if ponding water exists?
15. What are the warranty exclusions?
16. How thick is the membrane?
17. How chemically-resistant is the membrane?
18. How well does the membrane weather?
19. Does the membrane contain plasticizers that may be susceptible to ultraviolet exposure?
20. What is the dimensional stability of the membrane; is it prone to excessive shrinkage or contraction?

Installation

After answering these 20 questions, and selecting an SPR system, the designer's next task is monitoring the installation to ensure that the roof is installed correctly. Observing the contractor's approach to the following practices should give the designer some indication of the installer's professionalism.

Rolls of SPR membrane should be placed on the roof (over structural supports) as soon as the deck structure is in place. The SPR membrane should not be laid out on the ground. The rolls should not be concentrated in one area of the roof, but distributed immediately and evenly to avoid overloading the structure. Roof insulation should be placed on pallets not more than two tiers high, and covered with secured tarpaulins.

The roof area should be laid out in advance to utilize the largest sheets possible. This approach to layout should be carried out in advance by all bidders, who should include in their submittals to the architect a roof sheet layout, showing seams and penetrations.

Before the membrane is unrolled, the roof surface should be checked to ensure that there are no sharp objects, such as gravel or bits of metal, that might pierce the membrane. After the sheets are unrolled and positioned, seam splicing begins. There is little margin for error in splicing seams, since SPR offers no second or third ply of membrane to back up the first in the event of a leak. A manufacturer's representative should be on site during this phase, at least initially, to assure the quality of the splices. To make a field seam in an EPDM roof, the top sheet should be folded back 12 inches, showing at least the same amount of membrane of the adjoining sheet. This 24-inch wide area should be wiped clean of talc and debris, using solvent or cleaner furnished by the manufacturer. Inadequately cleaned splices are a common cause of seam failure.

The splice cement should be labeled by the manufacturer of the membrane and thoroughly mixed using a drill-activated stirring device. Most manufacturers' specifications call for a three-to-four inch minimum splice. However, most professional roofing contractors generally require about six inches. The extra cost of the materials is negligible, and the added splice area strengthens this critical juncture. One manufacturer specifies an in-the-splice sealant for all of its mechanically attached systems. Most manufacturers specify that adhesive be brush-applied, but most roofing contractors prefer a four-inch long-handled roller. The object is to achieve a thin, even film of adhesive on both mating surfaces; too much adhesive distorts the membrane. When the adhesive becomes tacky, but is not yet dry, the installer should carefully fold the top sheet over, and brush toward the seam with a roller or the palm of the hand, avoiding wrinkles or "fishmouths". If fishmouths do occur, they should not be "pressed down," but cut out. A patch should then be applied over the area of the cut-out, in accordance with the manufacturer's directions. The splice should then be rolled toward the seam with a heavy metal hand roller.

After the joint is cleaned, a bead of lap sealer should be applied along the exposed edge. It is important that this be done at the end of each day's work to prevent moisture from contaminating the seam adhesive overnight. The contractor should furnish the owner with an "as built" drawing showing the location of every seam.

Attachment Methods

There are four methods of attaching the SPR membrane to the substrate or deck: using gravel slag ballast, mechanically fastening, partially adhering using "spots" or "lines" of adhesive, and fully adhering. As illustrated in Figures 8.3a through d, many of the SPR products may utilize any of the attachment methods.

Mechanical attachment can present certain problems, the most common of which are corrosion of the fasteners and thermal bridging. Fasteners are normally required to provide a minimum of 300 pounds of pull-out resistance. In steel decks of less than 22-gauge thickness, and in lightweight concrete and cementitious decks, relatively expensive fasteners must be used. Another problem of mechanically attached membranes is wind-induced *flutter* around the fasteners, which may lead to fastener or membrane failure. On high-rise buildings, or in very windy locations, it is prudent to use fully-adhered systems. Many manufacturers are currently developing dependable mechanical fastener systems and adhered systems, in response to a trend away from ballasted systems. Many experts agree that there is no system on the market with a sufficient "track record" to predominate.

New Developments in SPR Systems and Attachments

Manufacturers continue to develop new methods for fastening and adhering SPR systems. This entire field of design and product development is changing so quickly that the average building designer or roofing specifier cannot hope to stay up to date on all of the latest products offered.

As of this writing, one manufacturer is now marketing a system of "non-attached" attachment. This method involves a series of vents that create a negative pressure beneath the roof membrane. This negative pressure increases in proportion to the wind velocity: the stronger the wind speed, the more tenaciously the membrane adheres to the deck. This system, which requires some traditional adhesion at the building perimenter, is said to relieve the installer of the expensive and time-consuming task of adhering, mechanically attaching, or ballasting the roof. If proven effective, this development may result in even more change in the SPR field.

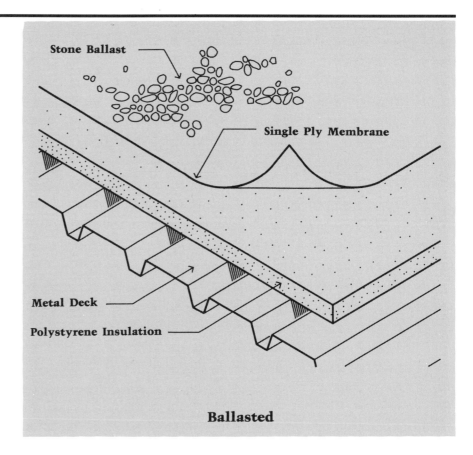

Figure 8.3a

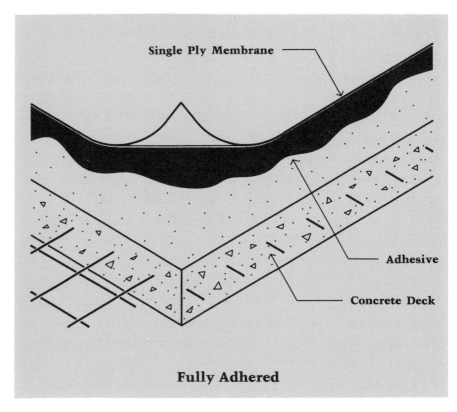

Figure 8.3b

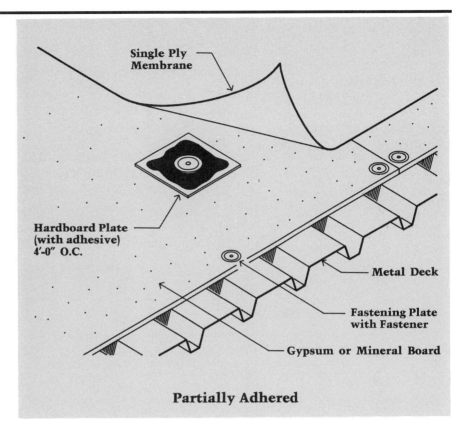

Figure 8.3c

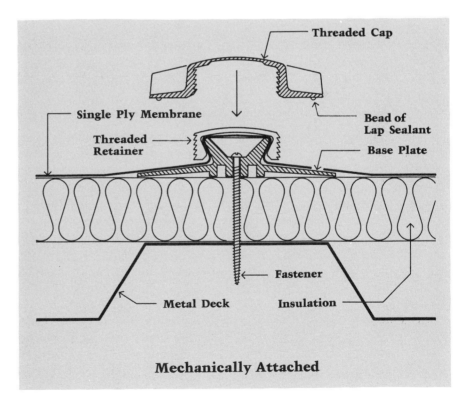

Figure 8.3d

Single-ply roofing is a relatively recent development in the industry. Nevertheless, after extensive research and development, SPR has taken its place as one of the major systems used in commercial construction today.

Summary

Single-ply offers the advantages of greater flexibility, moisture resistance (provided it is not torn or punctured), ease of repair, and a cleaner installation process. It also has excellent resistance to heat, ozone, and ultraviolet light. The selection process must be conducted carefully in order to avoid chemical incompatibilities and installation difficulties, and to choose the appropriate material for the area, color, and expected traffic.

CHAPTER NINE

METAL ROOFING

CHAPTER NINE

METAL ROOFING

Pre-engineered Structures

The use of corrugated iron as a building material dates back several centuries. For over 200 years, it has been a popular roofing material, following the use of grass thatch, stone, slate, tile, and lead.

Metal roofing is relatively lightweight, and has strong architectural appeal and durability. It is relatively easy and quick to install, and is readily available in many colors, textures, and profiles. For all these reasons, metal roofing continues to be a major factor in the roofing material market.

Almost half of all roofs constructed today in the low-rise, non-residential market are metal roofs — principally pre-engineered metal roofs. Most of these roofs are installed by steel erectors, not roofing contractors. The pre-engineered metal roof is comprised of light gauge (24- or 26-gauge) ribbed roof panels set on wide-spaced purlins. The most common material is galvanized steel, occasionally with a colored, baked enamel finish. There are also numerous aluminum systems on the market.

Types of Pre-engineered Metal Roofs

The two most commonly used types of pre-engineered metal roof systems are:

1. Screw-attached systems using fastener heads sealed with neoprene washers (see Figure 9.1)
2. Systems with high locking ribs and concealed fastener clips (see Figure 9.2)

The advantages of the first type of roof, screw-attached systems, are speed and simplicity of installation. Both the material and labor are less expensive for this type of roofing than they are for the concealed fastener systems.

The best pre-engineered systems on today's market are those in the second category, which fasten the roof panels to clips, rather than directly to framing members. In this way, the roof is permitted to "float," thereby providing for the natural expansion and contraction of roof panels without damage to the seams or fasteners. While this high-ribbed roofing using expansion cleats and no transverse joints is more expensive, its benefits are notable.

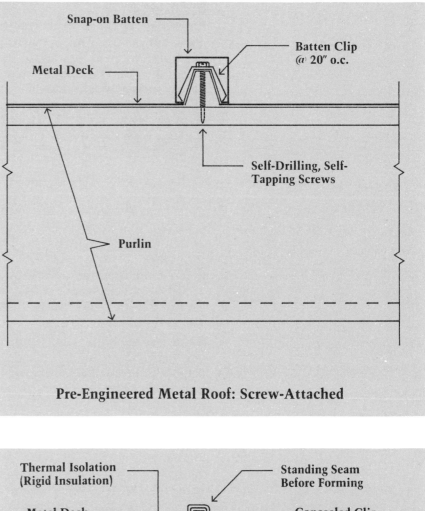

Pre-Engineered Metal Roof: Screw-Attached

Figure 9.1

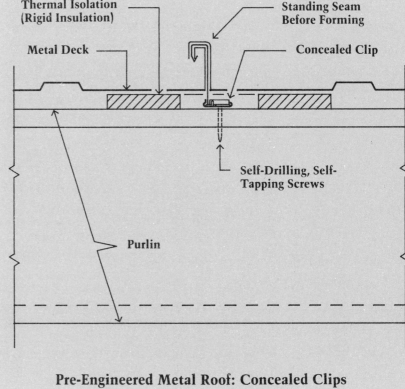

Pre-Engineered Metal Roof: Concealed Clips

Figure 9.2

Flashings

Flashings are more difficult to seal for metal roofs than they are for other types of roofing because of the configuration of the roof and the light gauge of the metal. Penetrations such as skylights, ventilators, and air conditioning units call for special detailing and neoprene sealer strips (see Figure 9.3). Hips, valleys, parapets, and penetrations should be eliminated from the design of metal roofs wherever possible. It is also important that all flashing be designed to be at least eight inches above the level of the roof to prevent water intrusion over the flashing during heavy rain or snow.

Maintenance and Repairs

Although they are inexpensive, neoprene-gasketed, screw-fastened roofs offer few good repair solutions. Maintenance personnel and building owners are being confronted with the problem of leaks which are difficult to repair. One source of trouble is traffic and movement of equipment across these roofs, helping to create depressions and buckled seams. Cementing or siliconing the screw heads stops minor leaks temporarily, but water ponding in the area eventually destroys the seal, allowing leaks to return. Another source of leaks is hardening and cracking of the neoprene fastener gaskets.

A "patch" made of glass mesh and asphalt cement may be used to repair isolated leaks. For more extensive leaks, an elastomeric coating may be applied to the surface of the roof, with reinforcing mesh at critical areas. There are several manufacturers of these water-based coatings, which are designed to be applied like paint to a thickness of approximately 35 dry mils. The problem is that because of the roof's ribbed surface, it is very difficult to control the thickness of the coating. The coating material normally has a five-year limited warranty. The cost of elastomeric coatings depends on the size of the roof and the location.

If roof leakage and damage extends beyond moderate repair using methods such as those described above, one might consider either installing a single-ply roof (SPR) over the old metal roof, or removing the old roof and replacing it with a new system. The first of these options, the single-ply roof, accommodates the building movement, thereby eliminating many sources of roof damage. It can be adhered or mechanically attached to the original metal roofing. Before the SPR is installed, the spaces between the ribs should be filled with rigid insulation board and a wide nailer installed at the perimeters (see Figure 9.4). Air handling units and other equipment should be raised up on supports that can be properly flashed.

Most manufacturers require that if the deck is lighter than 22-gauge, the fasteners must withstand a pull-out test of over 350 pounds per fastener. Most existing pre-engineered systems fail this test, making it necessary to use toggle bolts as an additional fastening device. Many contractors install a $1/2''$ plywood deck over the light metal deck before installing the new SPR system.

Installing a single-ply roof over a failing metal roof is an option in cases of extensive roof leakage, but investing in such a repair should be weighed against the cost and benefits of total roof removal and replacement.

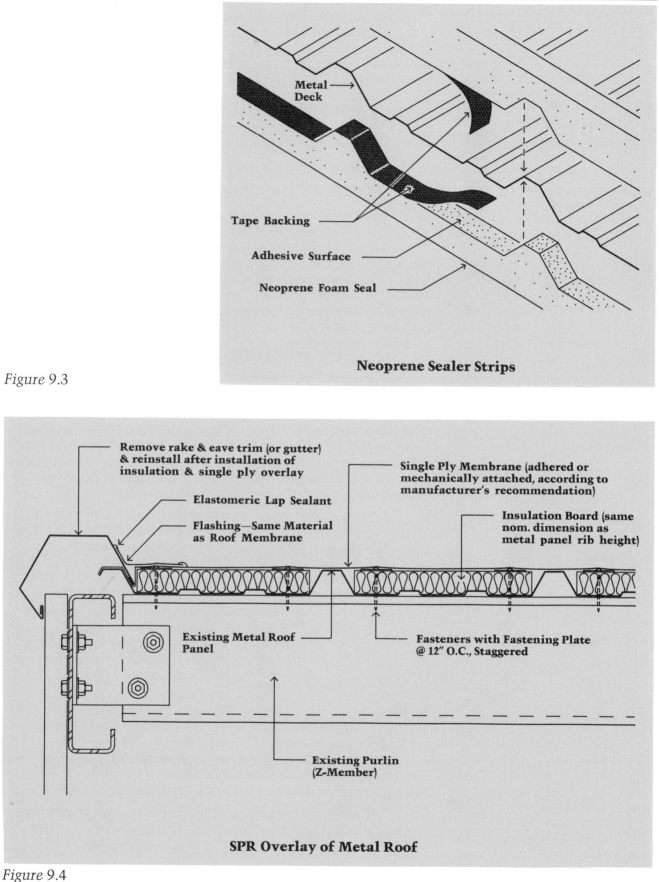

Metal Deck

Tape Backing

Adhesive Surface

Neoprene Foam Seal

Neoprene Sealer Strips

Figure 9.3

Remove rake & eave trim (or gutter) & reinstall after installation of insulation & single ply overlay

Elastomeric Lap Sealant

Flashing—Same Material as Roof Membrane

Single Ply Membrane (adhered or mechanically attached, according to manufacturer's recommendation)

Insulation Board (same nom. dimension as metal panel rib height)

Existing Metal Roof Panel

Fasteners with Fastening Plate @ 12" O.C., Staggered

Existing Purlin (Z-Member)

SPR Overlay of Metal Roof

Figure 9.4

Custom-fabricated Metal Roofs

Common Materials

Custom metal roofs are the most visually appealing and versatile types of systems for use on custom designs or in achieving intricate details, such as barrel roofs and convex mansards. Commonly used materials are standing seam and batten seam terneplate, copper, or galvanized steel.

Terneplate: Terneplate is prime (highest quality), copper-bearing steel, coated on both sides with an alloy of 80% lead and 20% tin. Developed in Wales in the early part of the nineteenth century, terneplate was used for roofs on many of the early structures in the United States, and is still used today in rehabilitations, additions, and period architecture. It is currently manufactured in the United States and is available in several gauges. It is most commonly supplied in 28-gauge with rust-preventative coating.

A terneplate roof should not be installed over structures with a slope of less than three inches to the foot. The deck should be stable and of a material capable of receiving and holding nails. Plywood is the preferred deck material and should be a minimum of $5/8''$ thick. Tongue-and-groove wood sheathing decks are also commonly used. The deck should be covered with a #30 asphalt felt. The felt should be covered with a red resin paper slip sheet to avoid bonding of the felt to the metal.

The main disadvantage of terneplate roofing is its high cost. Because installation is labor intensive, the relative cost for labor is often prohibitive. A terneplate roof requires greater skill for both installation and repair. Also, unless it is terne coated stainless, a terneplate roof must be painted to protect it from rust.

The cost of installing terneplate has been somewhat reduced by the use of roll-forming equipment for fabricating continuous pans and creating the double lock joint. Figure 9.5a illustrates conventional standing seam construction with a transverse joint. Figure 9.5b shows how the transverse joint can be omitted by using the expansion cleat in lieu of the traditional double cleat. This innovation can also reduce unsightly "oil canning" (wrinkling), which has been a problem with terneplate roofing.

Steel has a relatively low coefficient of expansion and contraction, which makes it more appropriate for these kinds of installations than aluminum. With these cost reduction features, the formed-in-place, "traditional" standing seam roof is again an option for the designer.

Other Custom Metal Roof Materials: Other materials commonly used in custom-fabricated roof systems are 16 oz. copper and 24 gauge galvanized roll stock. These materials are used in standing seam and batten seam systems, with many different seam details, one of which is illustrated in Figure 9.6.

Pre-formed Metal Roofing

Between the relatively inexpensive pre-engineered metal building systems and the custom-designed, site-formed metal system is the largest sector of the metal roofing market — pre-formed, pre-painted metal roofing.

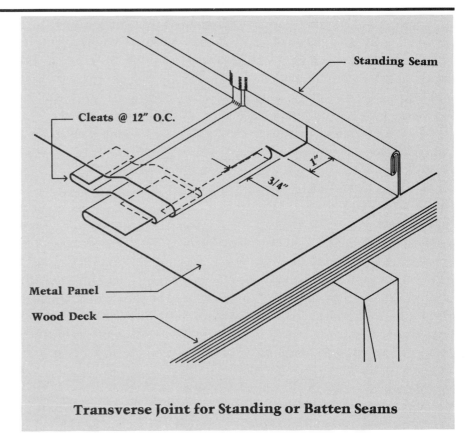

Standing Seam

Cleats @ 12" O.C.

1"

3/4"

Metal Panel

Wood Deck

Transverse Joint for Standing or Batten Seams

Figure 9.5a

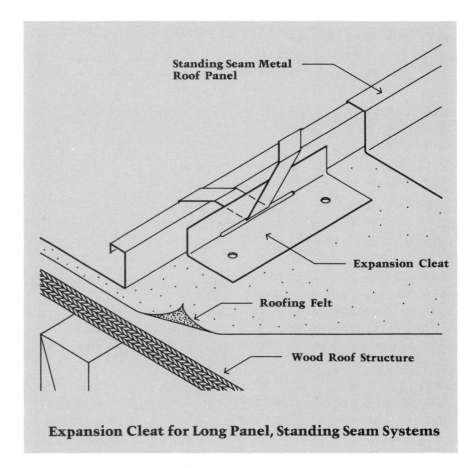

Standing Seam Metal
Roof Panel

Expansion Cleat

Roofing Felt

Wood Roof Structure

Expansion Cleat for Long Panel, Standing Seam Systems

Figure 9.5b

New Technology

Advances in roofing technology have made available a wide variety of panel profiles and colors in permanent finishes such as Kynar 500 and fluorocarbon with preapplied seam sealers.

However, the greatest strides in metal roofing technology in the past decade have been in the development of numerous "concealed clip" anchoring systems, as mentioned previously. (See Figure 9.6 for one example.) These clips are a boon to designers, although they can present problems if improperly installed. Correct alignment is critical in order to prevent binding between a panel and the clip(s) that retain it. When properly installed, these clips allow the system to meet rigid wind uplift standards, while allowing considerable movement in the "horizontal" plane, without the attendant leaks caused by protruding fastener heads.

Installation

There are two predominant methods of joining pre-formed metal roof panels: *snap-on seam clips* (see Figures 9.7a and b), and *automated seaming* using electric field-seaming machines. The seaming operation locks concealed, sliding clip fasteners into place, forms the seam, and can even apply joint sealant as a part of the operation.

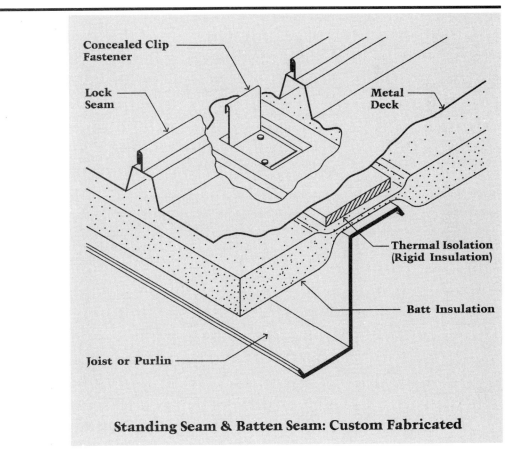

Figure 9.6

Standing Seam & Batten Seam: Custom Fabricated

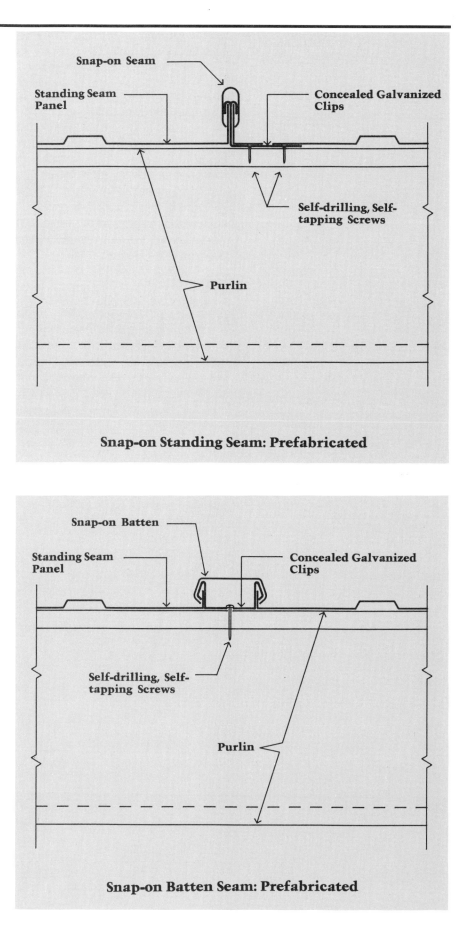

Figure 9.7a

Snap-on Standing Seam: Prefabricated

Figure 9.7b

Snap-on Batten Seam: Prefabricated

Summary Metal roofing, a popular material for over 200 years, offers particular advantages to today's owners and roofing contractors. New technology in pre-engineered, custom-fabricated, and pre-painted metal roofing systems provides a variety of solutions for many design and construction requirements.

CHAPTER TEN

SPRAYED-IN-PLACE FOAM

CHAPTER TEN

SPRAYED-IN-PLACE FOAM

Polyurethane foam (PUF) is not a new material; the chemical reaction of a polyol and an isocyanate to create polyurethane foams was discovered 50 years ago by Dr. Otto Bayer. In combination with various catalysts, surfactants and blowing agents, this family of materials has been used for many years to insulate tanks, refrigeration, and piping.

Uses and Application Methods

Applied using one of four basic methods, PUF is well suited to a number of purposes. The *pour* and *froth* systems are used in cavity-filling applications. *Board stock* is factory-formed, and when faced with different materials, is used in buildings, freezer and cooler construction, and other insulating roles, including roof insulation and membrane underlayment.

In the *sprayed-in-place* system, PUF is applied with a special spray gun which mixes the "Part A" and "Part B" components and deposits the foaming mixture onto the roof surface. When properly applied, PUF forms a rigid, homogeneous "single ply," which is then topped with a coating to protect the thick "membrane" from ultraviolet light, moisture, and impact or traffic damage.

Early Problems and Their Solutions

In the first several years after PUF was introduced, it proved to have certain drawbacks. Once in place, it suffered from various failures caused by: improper adhesion to the substrate; UV degradation; deflection of the deck, causing cracking of the PUF; water contamination; shrinkage of the material, causing cracking and delamination; and uneven application thickness, resulting in ponding.

Thanks in large part to the efforts of the Polyurethane Foam Contractors Division (PFCD) of the Society of the Plastics Industry (SPI) in Washington, D.C., spray-applied PUF has come a long way in the past few years. Large, research-based manufacturers became involved in the testing and improvement of PUF, and advancements were made. As a result, PUF has become more of an option in the new roofing, and especially the re-roofing, market.

Application Guidelines

Adhesion
Before applying the PUF, a primer is required in many cases in order to attain proper adhesion to the substrate. Individual manufacturers are the best source of information on these

requirements. *Lightweight or insulating concretes are not recommended for PUF application.* The PUF layer must be adequately adhered to the substrate in order to resist wind uplift and movement. When using a vapor retarder underneath the PUF, the installer must again follow the manufacturer's recommendations to assure adhesion.

Preparation

All deck debris, dirt, gravel and contaminants must be removed prior to the placement of PUF. Care should be taken to ensure that all spans of deck are designed for a maximum deflection of $1/240$th of the span. All deck joints with an opening greater than $1/4''$ should be sealed prior to applying PUF. Unsupported joints in metal deck should be lapped properly and fastened. The deck should be dry and free of oil, grease, or other contaminants. The flutes of metal decks should either be filled with filler material, or covered with substrate board in a manner that meets Factory Mutual criteria. Lightning rod cables should not be embedded in PUF; they should instead be located above, and isolated from the roof surface. Lightning rods should be masked prior to foaming. Electrical conduit should likewise never be embedded in the PUF, but instead located within the building attic area. Wood plank deck, even if tongue-and-groove, should be overlaid with a minimum of $1/4''$ exterior grade plywood or other suitable substrate.

Thickness Control

The minimum recommended thickness for each "pass" of foam sprayed in place is $1/2''$. The minimum total PUF thickness is one inch. If necessary to provide adequate drainage, the PUF may be tapered around the drains. However, this practice is discouraged. In no case should the material be tapered to a thickness of under one inch, except at feathering for flashings (see Figure 10.1), drains (see Figure 10.2), and other such terminations. The uniformity of thickness and the surface finish required for a lasting, trouble-free installation cannot be obtained by amateur installers. Although advances are being made in the area of robotics and mechanical application methods, manual application is still the most common method. Experience and skill are required to achieve a uniform pass thickness as well as acceptable edge-feathering for self-flashing and other terminations.

Surface Quality

Appropriate application techniques, equipment maintenance, and favorable weather conditions should result in the desired surface texture. Proper surface texture descriptions and illustrations are outlined below and in Figure 10.3. Some surface imperfections are normal. For example, *smooth, orange peel* or even *verge-of-popcorn* surface, if properly coated with protective coating, are all acceptable. However, *popcorn* or *tree bark* texture may be unacceptable, in which case they must be removed and refoamed to an acceptable finish. It is imperative that installers achieve a texture acceptable to the manufacturer's field representative; approval of these representatives is needed in order to obtain the manufacturer's certification and warranty of the system.

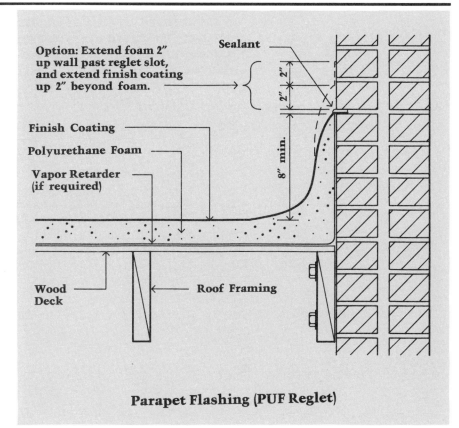

Parapet Flashing (PUF Reglet)

Figure 10.1

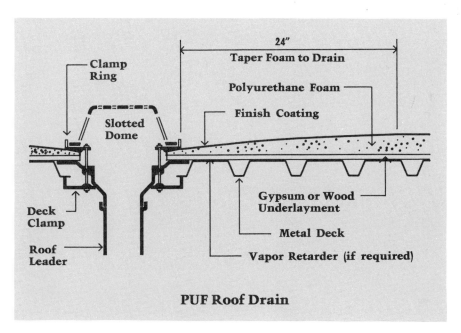

PUF Roof Drain

Figure 10.2

The surface quality of the foam has a significant impact on the quality of the system. If the surface texture is either *popcorn* or *tree bark* (see Figure 10.3), it is difficult to apply a uniform coating. If there are pinholes or voids, moisture can work its way to the foam and destroy the bond between the coating and the foam. If the foam is exposed to ultraviolet light, rapid degradation accelerates the absorption of moisture and damage to the roof. For this reason, PUF has had rather limited acceptance.

Expansion and Control Joints

Urethane foam, with its high coefficient of thermal expansion, may require more expansion joints, as shown in Figure 10.4, or more control joints according to the published installation guide of the particular PUF manufacturer. These joints should be placed in both directions at all re-entrant corners, to form rectangular "pours". Again, installers should refer to individual manufacturer's design data for recommended joint spacing.

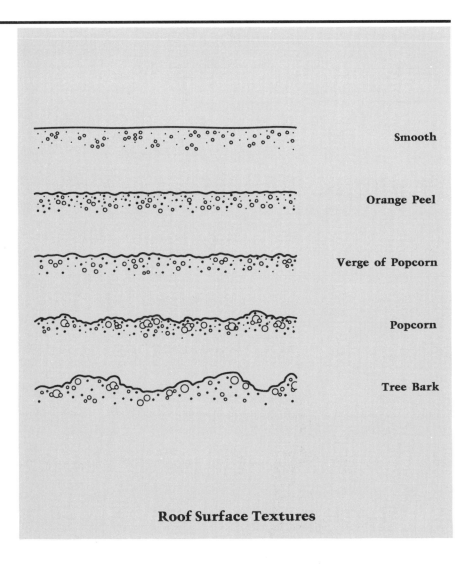

Smooth

Orange Peel

Verge of Popcorn

Popcorn

Tree Bark

Roof Surface Textures

Figure 10.3

Accessories

Tapered joint and penetration "flashing" is used at all accessories, discontinuities, and joints, with the elastomeric coating system (applied in several thin coats) serving the actual role of flashing. Eave flashing, fascia strip, or *foam stop*, as it is referred to in PUF terminology, is a critical element in the PUF system. Consequently, it should be detailed as in Figure 10.5, with the roof sloping away from the outside edge, and towards interior drains. This arrangement helps to keep water from ponding at the junction of foam stop and PUF, where, over time, it might otherwise seep into any discontinuity in the protective coating.

Protective Coating

The elastomeric coatings used on PUF systems serve to protect the polyurethane from UV degradation. It is imperative that the coating be applied on the same day as the PUF, or at least within 24 hours. If moisture should enter the PUF before the coating is applied, severe degradation may occur. If the PUF attains a tan or cream colored dusty surface, it may have undergone excessive degradation, and it may be too late to obtain protection with an application of coating.

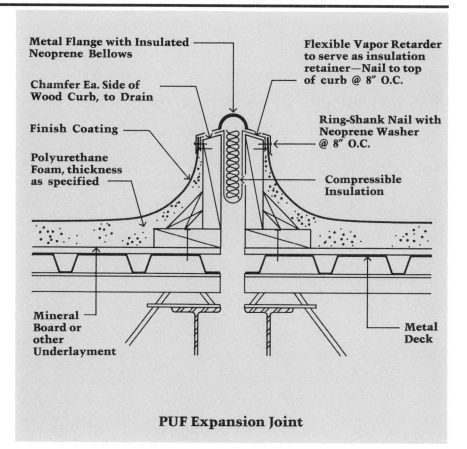

PUF Expansion Joint

Figure 10.4

Coating Classifications: Elastomeric coatings, by definition, are capable of 100% elongation under load, and can recover to their original dimensions. They may be classified according to the following types:
- Acrylic Latex
- Urethane
- Silicone
- Neoprene (Hypalon)
- Butyl Rubber
- Epoxies
- Modified Asphalts

It is important to be sure that the coating used is compatible with the PUF product, and approved by the PUF manufacturer.

Application of Coatings

Primer Requirements: Some coatings, such as acrylic latex, vinyls, and asphalts require that PUF first be sprayed with a primer for proper adhesion. When in doubt, consult manufacturers' recommendations for primer requirements.

Coverage: Virtually all elastomeric coatings for PUF systems require two or more coats for proper coverage. Each coat is applied at right angles to the preceding coat. Alternating coats should be in contrasting colors. This system provides ready evidence of incomplete or thin coats in most systems.

Each type of coating has a corresponding minimum Dry Film Thickness (DFT), which varies from product to product from 10

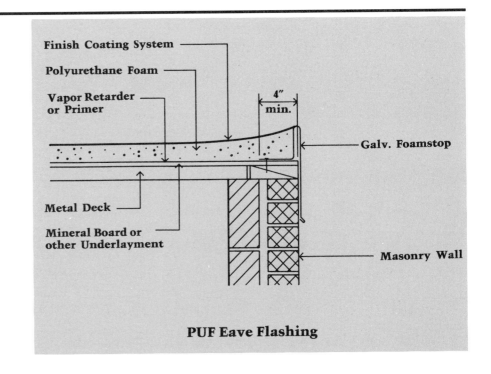

PUF Eave Flashing

Figure 10.5

mils to 30 mils. To achieve these varying film thicknesses, different amounts of product must be sprayed on the hardened PUF surface. The amount of product depends on the PUF surface texture. Figure 10.6 is a chart listing and describing the various states from smooth to rough of the PUF surface. The chart also provides recommended quantities of coating material for each acceptable surface texture. Also included is a formula for calculating required coating quantities and an example calculation for a hypothetical project.

Coating Quantity Required

Surface Texture Description	Surface Characteristics	Additonal Material Percentage (AMP)*
"Smooth"	Shows spray undulation, but no pronounced nodules or valleys	.05
"Orange Peel"	Shows a fine texture, similar to that of an orange peel.	.10
"Verge of Popcorn"	The roughest acceptable texture suitable for receiving coating. Nodules are larger than valleys; valleys are curved, not sharp.	.50
"Popcorn"	Large nodules on surface; valleys form sharp angles. Unacceptable surface for coating.	n/a
"Treebark"	Large nodules and ridges; valleys form sharp angles. Unacceptable for coating.	n/a

*the additional material percentage (AMP) varies with the roughness of the surface texture.

The following formula may be used to calculate the actual amount of coating material required for a particular project:

Gallons of coating per 100 ft² roof surface = .0623 x (1+Amp) x (DFT/SCV)

Where: AMP = Add'l. Material Percentage (table)
DFT = Dry Film Thickness required on job (specifications)
SCV = Solid Content by Volume (indicated on material container)

Example: To coat a 100,000 sq. ft. roof that exhibits a fine, *orange peel* texture, the installer selects a coating with a 60% solid content by volume—and specifies a 30 dry mil. thickness for the coating.

Gal./100ft² = .0623 x (1 + .10) x (30/.60)
Gal./100 ft² = 3.427

For a 100,000 ft² application, 3427 gallons are required. However, it would be close enough to call for 62 drums of coating, at 55 gallons each — a total of 3410 gallons.

Figure 10.6

Summary   For many large re-roofing projects with unusual surface irregularities or shapes, or even for conventional projects, PUF has proven to be a cost-effective material with many attributes. The early problems in the field, and even those that continue to occur, are not indicative of the material's inherent shortcomings, but rather are the result of the inexperience of many installers who attempt application without proper training. Standards and testing for sprayed polyurethane foam continue at ASTM, NIST, NRCA, and PFCD (SPI), as well as in the research laboratories of corporations in the U.S. and the National Research Council of Canada. (See Appendix A for a listing of industry organizations for further information on PUF systems, suppliers, and guide specifications.)

CHAPTER ELEVEN

SLATE AND TILE ROOFING

CHAPTER ELEVEN

SLATE AND TILE ROOFING

Slate Slate has long been a popular roofing material, partly because of an inherent property that allows it to be easily split to fit size and shape requirements. A further advantage of slate is that it is practically non-absorbent. The origin of a particular slate determines its color and quality. There are active slate quarries in Maine, New York, Pennsylvania, Vermont, and Virginia. Blue-gray slate generally comes from Pennsylvania, green or purple from Vermont, and red from New York.

The most common thickness used for roofing slate is $^3/_{16}$″. Figure 11.1 shows the available sizes of standard slate, the minimum number of slates required per square, the exposures, and the quantity (by weight) of nails required per square of roof surface. Copper nails are the only suitable nails for use with slate. Nail length may be determined by adding one inch to twice the thickness of the slate. Nails should be large head, diamond-pointed, #10 gauge shank.

Weight of Slate
Depending on the thickness of the material chosen, slate can weigh anywhere from 650 to over 750 lbs. per square (100 sq. feet) of roof area, including required head laps. When the weight of nails, felt, and standard $^3/_{16}$″ slate is added, the dead load of slate roofing is roughly 800 lbs. per square. For $^1/_4$″ material, approximately 1050 lbs. per square should be used, and for $^3/_8$″ slate, about 1550 to 1660 lbs. per square.

Imitation Slate
There are several manufacturers of imitation slate, a compound made of mineral fiber and cement. Advertised as having the appearance and durability of real slate, at about 50 percent of the cost of a comparable slate roof, these products have found a considerable market. The disadvantages of this material have been fungal growth and loss of colorfastness due to its porosity.

Traditional Installation Practices
Because of the relatively high cost of slate, the market for this material has decreased. As a result, fewer young roofing contractors are going into this roofing specialty. With insufficient volume to justify training programs, the skills required for slate

roofing will have to be passed on from older to younger tradesmen on the job site. Some basic guidelines for slate roof installation are as follows. See Figure 11.2 for an illustration of a slate roof system.

- No through joints should occur from the roof surface to the felt.
- The overlapping slate should be joined as near the center of the underlying slate as possible, and not less than three inches from any underlying joint.
- A standard three inch *headlap* should be used. (Headlap is that portion of the slate which overlaps the lowest slate course beneath it.)
- The exposure (weathering surface) should be determined by subtracting three inches from the length of the slate and dividing by two.

Schedule For Standard $^3/_{16}$″ Thick Slate

Size of Slate (In.)	Slates Per Square	Exposure with 3″ Lap	Nails Per Square (lbs.)	Nails Per Square (ozs.)	Size of Slate (In.)	Slates Per Square	Exposure with 3″ Lap	Nails Per Square (lbs.)	Nails Per Square (ozs.)
26x14	89	11$^1/_2$″	1	0	16x14	160	6$^1/_2$″	1	13
					16x12	184	6$^1/_2$″	2	2
24x16	86	10$^1/_2$″	1	0	16x11	201	6$^1/_2$″	2	5
24x14	98	10$^1/_2$″	1	2	16x10	222	6$^1/_2$″	2	8
24x13	106	10$^1/_2$″	1	3	16x9	246	6$^1/_2$″	2	13
24x11	125	10$^1/_2$″	1	7	16x8	277	6$^1/_2$″	3	2
24x12	114	10$^1/_2$″	1	5					
					14x12	218	5$^1/_2$″	2	8
22x14	108	9$^1/_2$″	1	4	14x11	238	5$^1/_2$″	2	11
22x13	117	9$^1/_2$″	1	5	14x10	261	5$^1/_2$″	3	3
22x12	126	9$^1/_2$″	1	7	14x9	291	5$^1/_2$″	3	5
22x11	138	9$^1/_2$″	1	9	14x8	327	5$^1/_2$″	3	12
22x10	152	9$^1/_2$″	1	12	14x7	374	5$^1/_2$″	4	4
20x14	121	8$^1/_2$″	1	6	12x10	320	4$^1/_2$″	3	10
20x13	132	8$^1/_2$″	1	8	12x9	355	4$^1/_2$″	4	1
20x12	141	8$^1/_2$″	1	10	12x8	400	4$^1/_2$″	4	9
20x11	154	8$^1/_2$″	1	12	12x7	457	4$^1/_2$″	5	3
20x10	170	8$^1/_2$″	1	15	12x6	533	4$^1/_2$″	6	1
20x9	189	8$^1/_2$″	2	3					
					11x8	450	4″	5	2
18x14	137	7$^1/_2$″	1	9	11x7	515	4″	5	14
18x13	148	7$^1/_2$″	1	11					
18x12	160	7$^1/_2$″	1	13	10x8	515	3$^1/_2$″	5	14
18x11	175	7$^1/_2$″	2	0	10x7	588	3$^1/_2$″	7	4
18x10	192	7$^1/_2$″	2	3	10x6	686	3$^1/_2$″	7	13
18x9	213	7$^1/_2$″	2	7					

TABLE 1
Schedule for Standard Slate

Figure 11.1

Ridge Construction: Several methods may be used to construct the ridge on a slate roof, but the most common, shown in Figure 11.2, is called the *saddle ridge*. Using this method, the slates are brought together at the ridge so that the opposing slates will butt flush (see Figure 11.2). Wood lath is then nailed to the ridge so that the top row of slate will be properly aligned. To ensure that no moisture can be driven into the ridge, a strip of 60 mil. EPDM roof membrane is then placed over and along the ridge. The final *combing slate* is then placed over the top row of slate, with a three-inch overlap. The ridge course should be run with the "grain" horizontal. The ridge slates are secured to the deck with two nails, which are then covered with plastic cement. All nails, except the last one on each side of the ridge, are concealed with a cover of cement.

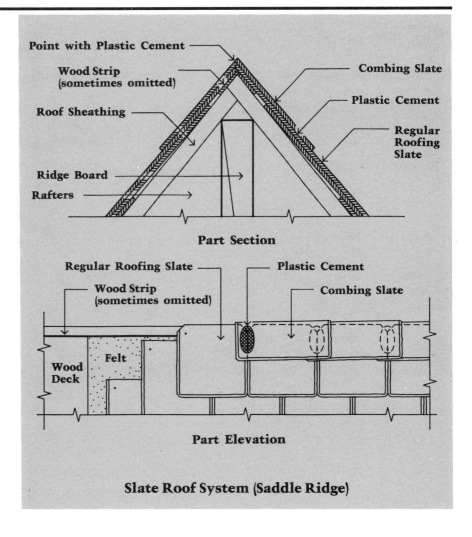

Figure 11.2

Hip Construction: There are also several methods that can be used to form the hips. Again, the most common is the *saddle hip* shown in Figure 11.3. Here, as in the ridge, a wooden lath strip should be run along both sides of the hip. The ridge slate (of the same size as the exposure in the roof slates) should be attached with four nails per slate, with a strip of EPDM membrane underneath as shown, to prevent moisture infiltration. All nail heads should be covered with plastic cement.

Valleys: Most valleys in slate roofs are "open", as shown in Figure 11.4a. However, slate may also have closed valleys (see Figure 11.4b). The valley is first lined with sheet metal (preferably 16 oz. cold-rolled copper sheet) so that water will be channeled between the slate on both sides. The width of this channel should uniformly increase, at the rate of one inch per eight feet, down the valley toward the eave in order to carry the additional volume of water.

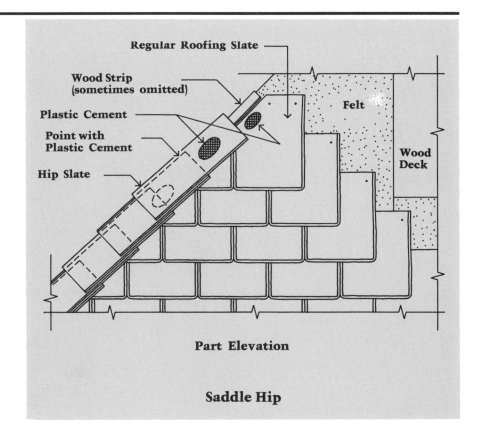

Figure 11.3

116

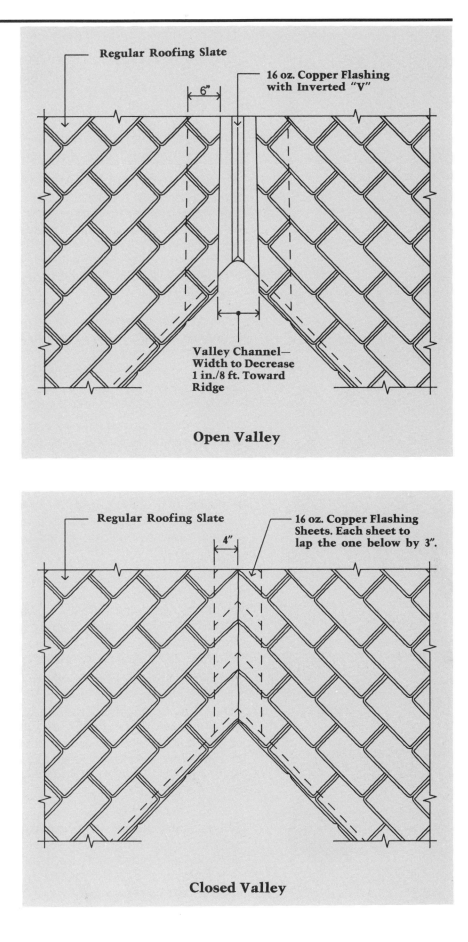

Figure 11.4a

Figure 11.4b

The slate application should begin two inches from the inverted "V" at the top and should taper away at the rate of one half inch for every eight feet down to the eave. The function of the inverted "V" in the valley is to "break up" the flow from each side and thereby prevent water from being driven up under the slate on the opposite side. The valley flashing should extend at least six inches under the slate and should be crimped back to intercept any moisture. The metal valley should be secured with cleats at eight inches on center, along both sides of the valley.

Maintenance and Repair

Since the cost of replacing a slate roof is often prohibitive, it is important that proper maintenance be performed in order to extend the life of the roof. The key element in an effective maintenance program is employing an experienced roofer to make all repairs. An inexperienced workman can destroy a slate roof. A visual inspection of the roof should be made with the mechanic on a regular basis.

Visual Inspection: If such inspection reveals any of the following conditions, a systematic approach to maintenance should be initiated.
- Many slates are missing or dislodged.
- The underlying felt or deck is showing.
- The metal flashing is missing or badly rusted.

Further examination may be conducted as follows: When tapped with a hard object, good slate emits a ringing sound. A dull sound means that water-damaged slate has absorbed too much moisture and has deteriorated. When slate begins to flake and come apart or becomes brittle and easily broken, it should be removed and replaced. It is usually not practical to completely remove a slate roof. Instead, only the damaged slates are removed, and the good ones are saved to be reinstalled along with the new slate.

Cost

Pure, unblemished gray is the most expensive type of slate. Because it is difficult to keep a roof perfectly clean, mottled gray or purple are the preferred colors. They can also be delivered at a significantly lower price. "Short", or smaller slate is less expensive than larger slate, but it requires more installation time.

Tile Roofing

Tile roofing is a versatile, almost "permanent" material. It was popular in ancient Greece where both flat tiles and curved tiles with joint covers were used, predominantly in shades of terra cotta. Both soft-formed and dry press-formed tiles were baked, like brick, in various sizes of firing kilns. In China and Japan, the craft of tile making was elevated to an art. Vivid colors and durable glazing were developed to yield a long-lasting, attractive material. Today, tile is seen throughout much of the U.S. in garden offices, commercial and retail buildings, restaurants, and residential buildings.

Tile is a very versatile roofing material — it is considered almost permanent; it is highly fire-resistant, exhibits a strong texture, and provides a structure with architectural interest. It should be noted that clay tile is a heavy material, and this type of roof requires appropriate structural allowances. Tile is also one of the most

118

expensive roof systems, along with slate and custom
-fabricated metals.

Tile Shapes

There are two general categories of clay tile — *roll tile* and *flat tile*.
Roll tile can be either flat with "pan" sides, or curved "barrel"-
shaped. It may have a smooth lap or interlocking lap edge. Some
examples of roll tile are shown in Figure 11.5. Flat tile has no
curved edges; it may lock together on its top and bottom edges
(classic interlocking), or on its sides, as in newer systems.
Alternately, flat tile may be laid shingle-fashion, as in the *Norman*
style. Figure 11.5 is an illustration of various title configurations.

There are hundreds of variations in tile shape and style within the
two general categories of flat and roll. Most clay tile produced in
the U.S. is manufactured in California, Florida, or Texas. The
largest producer of *glazed* tile roofing in the U.S. is located in Ohio.

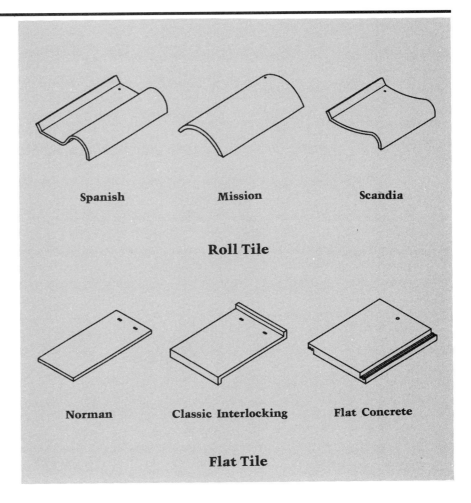

Figure 11.5

Ridge Tiles: Ridge tiles can be *closed-end, open-end* (gable), or *hipped* terminal tiles (See Figure 11.6). Each manufacturer specifies its own preferred method for lapping and sealing ridge tiles.

Eave and Gable Tiles: Special tile shapes are required for installation at eaves and gables, as well as beneath ridge tiles. Eave closures, top fixtures, and gable rake (see Figure 11.7) are used to finish out the installation of various roll tiles. Occasionally, metal closures are used at gables, eaves and ridges in lieu of these special tile fittings.

When flat tile is used, a *taper strip* or "under eaves" course should be used at the eaves to provide the proper angle for the first full course of tiles. This course should have a 3" separation every 48" to 60" to allow any trapped moisture to escape.

Required Slopes for Tile Roof

The slope of a tile roof should be no less than four-in-twelve. For low-sloped roofs (three inches to the foot), all configurations of clay tile may be installed *if* two layers of #43 base sheet are used; the first ply nailed and the second mopped in steep asphalt. Flat, shingle tile without an interlocking feature should not be installed on roof decks having a slope of less than 5" per foot. A minimum of one layer of #30 roofing felt should be applied as an underlayment. Since the underlayment is the waterproofing course, it is best to use a #43 base sheet for this purpose. While a high density clay tile roof will last several hundred years, most underlayment will not.

Ice Shields

For installations where the January mean temperature is below 30 degrees Fahrenheit, an ice shield consisting of #43 felt mopped in steep asphalt should be installed. The ice shield should extend from the eave to the inside of the wall line.

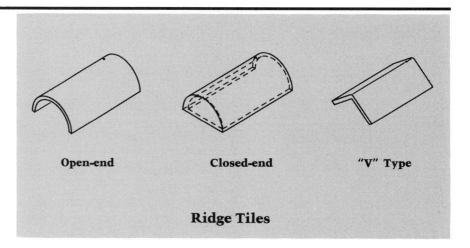

| Open-end | Closed-end | "V" Type |

Ridge Tiles

Figure 11.6

Valleys

Special attention must be paid to the manner in which open valleys are installed in tile roofs. The requirements include an extra ply of dry-in felt installed before the metal valleys are fitted. As shown in Figure 11.4, the metal valleys should be "V" crimped in the middle in order to lessen the force of the water running down the slope and prevent it from driving up under the tiles on the opposite side.

At the valleys, tiles must be cut along a line to form an even edge, using a power saw and carborundum blade. Metal valleys and flashing should always be fabricated out of copper, stainless steel, or other non-corrosive metal.

Handling Tiles and Avoiding Breakage

Tiles are susceptible to significant breakage in transit. They should, therefore, be handled as little as possible. Furthermore, tile should never be stored on the ground, but rather should be placed — on edge — on pallets or lumber blocking no more than four rows high. Tile should be mixed for color on the ground rather than on the roof in order to minimize the time that must be spent on the

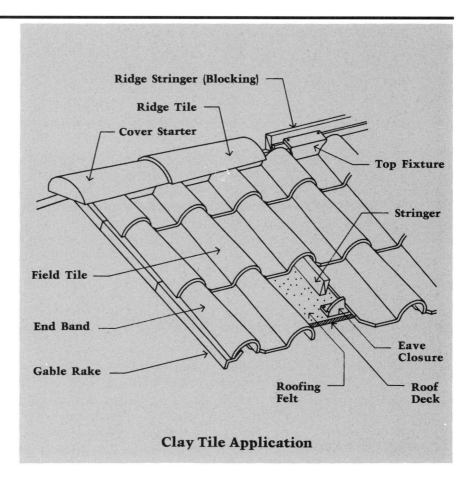

Clay Tile Application

Figure 11.7

roof. Tile manufacturers provide detailed installation instructions. When followed carefully, these guidelines are further protection against breakage. Instructions are normally included in each shipment or box of material.

Tile Installation Procedures

Decks: If plywood deck is used, the sheets should be separated by at least $^1/_{16}$". It should be exterior grade plywood of a thickness adequate to satisfy nailing requirements. If wood planking is used, it should be a minimum of 1 x 6 nominal lumber, spanning no more than 24 inches between rafters.

When the felt underlayment has been installed, vertical laths should then be nailed directly above all rafters, through the felt. Horizontal battens (stringers) are nailed across the lath, and spaced according to tile dimensions.

Stringers (Nailing strips): Hip and ridge stringers (See Figure 11.7) vary in height depending on the type of tile and the roof slope. It is best to lay out a few tiles at the ridge and determine the proper height for the ridge stringer.

Material Reuse

Tile labor installation costs vary significantly with the style of the tile, the size and nature of the installation, and the geographic location. Because of the high material cost, old tile roofs are normally repaired by carefully removing the old tile, replacing the deteriorated dry-in felts with #43 felt, and then reinstalling the tile, replacing only those tiles that are broken. This operation is very labor intensive, but in most cases is still cost effective. Depending on the regularity of the roof shape, the output of a roofer installing tile is generally between 35 and 75 square feet per hour. During this process, from 10-25 percent of the existing tile is likely to be broken and rendered unusable.

Note: For additional information on the use of tile roofing, contact the National Tile Roofing Manufacturers Association, 3127 Los Feliz Blvd., Los Angeles, CA 90039 (213) 660-4411.

Summary

Despite their relatively high cost, slate and tile are popular roofing materials in many parts of the country. Both are durable; in fact tile is considered almost permanent. Because of the high cost of these materials, any repair involves removal of the damaged tiles or slates, and salvaging those that are intact. This approach to repair is labor-intensive, but still cost-effective. Expensive repairs can be avoided or at least minimized by a program of regular inspection and maintenance by an experienced roofing tradesman.

CHAPTER TWELVE

SHINGLES AND SHAKES

CHAPTER TWELVE

SHINGLES AND SHAKES

Composition Shingle Roofs

Asphalt shingles were first manufactured in 1903 by Henry M. Reynolds in Grand Rapids, Michigan, who cut shingles from stone-surfaced sheets of roll roofing. Interestingly, he did not patent his invention. In 1914, F. C. Overbury of the Flintkote Company enhanced the asphalt shingles by applying crushed slate to help hold them in place during high winds. By the 1920's, this version of the asphalt shingle was widely used in the roofing industry. The product was relatively fireproof, easy to ship, and economical to install. The shingles were manufactured in a variety of shapes and colors. The shingle roof became "everyman's roof," and was even sold by mail-order catalog.

Three-tab Designs

By the 1950's, a predominant shingle design was emerging: a three-tab, thick-butt asphalt and felt shingle, weighing about 235 pounds per square. This type of shingle is illustrated in Figure 12.1. The three tabs were formed by *cut-outs,* each of which was often referred to as a *watercourse.* This watercourse between each tab improved the aesthetics of the shingle roof, but often proved to be the most vulnerable element, since this was where wind-induced stress concentrated, and water damage was first sustained. Shingle roofs always deteriorated first in the watercourses. Warranties of 15 to 20 years were offered. However, it was rare for any shingle roof to last the full term, unless installed on very steep slopes, where water contact duration and traffic were minimal.

In the 1970's, the manufacturers began to replace the organic core felt in the shingles with glass fiber felt. Theoretically, the glass fiber shingles had more resistance to wear. Unfortunately, at least in the earlier products, they were far more fragile than organic shingles. Problems occurred with delamination, splitting, and poor attachment.

After considerable product development, including an increase in product weight and lamination, glass fiber shingles now represent about 90 percent of the residential asphalt shingles used. This glass mat substrate is far superior in performance to the organic shingle substrate of paper (cellulose) fibers. Glass fiber shingles do not absorb moisture, so they retain dimensional stability. They do not rot, mold, or mildew. Glass fiber is incombustible; therefore, these

1	2	3			4		5	6
		Per Square			**Size**			
PRODUCT	Configuration	Approximate Shipping Weight	Shingles	Bundles	Width	Length	Exposure	Underwriters' Listing
Wood Appearance Strip Shingle More Than One Thickness Per Strip 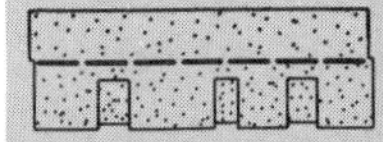 Laminated or Job Applied	Various Edge, Surface Texture & Application Treatments	285# to 390#	67 to 90	4 or 5	11$\frac{1}{2}$" to 15"	36" or 40"	4" to 6"	A or C — Many Wind Resistant
Self-Sealing Strip Shingle 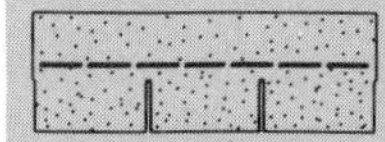	Conventional 3 Tab	205#- 240#	78 or 80	3	12" or 12$\frac{1}{4}$"	36"	5" or 5$\frac{1}{8}$"	A or C — All Wind Resistant
	2 or 4 tab	Various 215# to 325#	78 or 80	3 or 4	12" or 12$\frac{1}{4}$"	36"	5" or 5$\frac{1}{8}$"	

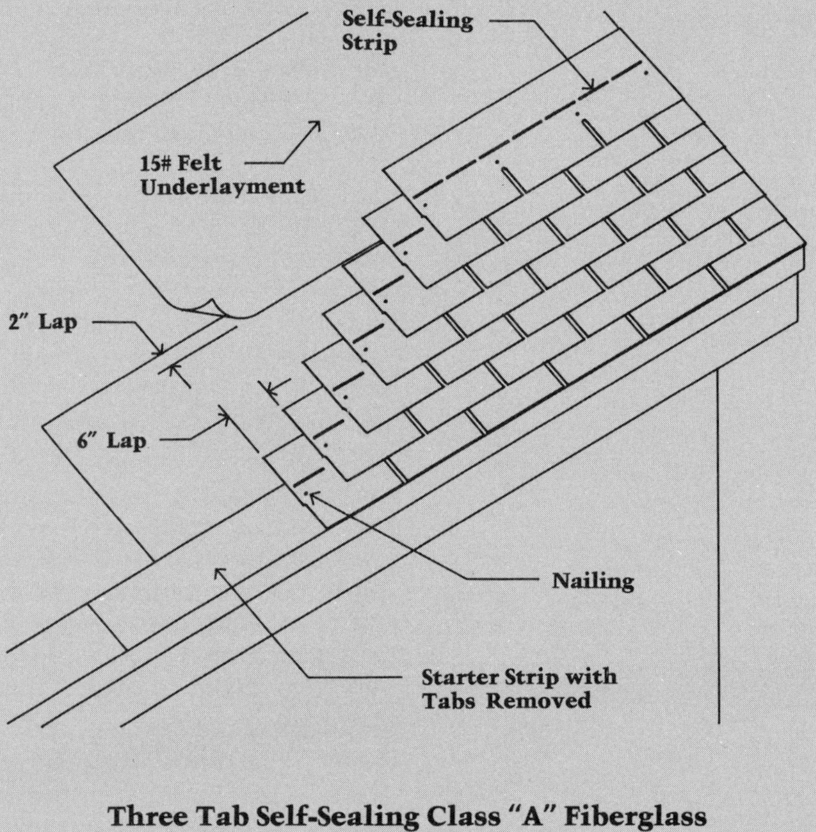

Three Tab Self-Sealing Class "A" Fiberglass Strip Shingles

Figure 12.1

shingles can carry a "Class A" fire rating as assigned by Underwriters' Laboratories. These shingles usually are covered by a 20-year warranty, and are readily available throughout the country.

Laminated Dimensional Shingles

For higher-budget applications, the *laminated dimensional shingle* is finding increased use. It does not have the cut-out of the three-tab shingle. It is thicker, due to a lamination of more than one thickness of product. It also contains shadow lines, resembling a wood shake or slate shingle.

Although little is said about shingle weight today, the most commonly used laminated shingles weigh from 285 to 390 pounds per square, as compared to 180 to 290 pounds for single-thickness shingles. There is little difference between the cost of labor for the application of lightweight versus laminated shingles, and material prices are growing more competitive. Therefore, it makes good sense to specify the laminated shingle when possible.

Installation Considerations

Shingles should be stored on pallets and covered until ready for installation. The bundles or pallets of shingles should be transported to the roof by conveyor or crane, not by hand. Bundles should be mixed to provide color blend.

During a "tear-off" of the old roofing materials, precautions should be taken to protect the building and grounds below. All old shingles should be removed, nails pulled, sheathing securely fastened, and rotten sheathing replaced. Although this step is commonly omitted, the #15 dry-in felt should be installed as shown in Figure 12.1. Roofing mechanics should strike lines at each course and install each shingle with four fasteners (also shown in Figure 12.1).

Flashing and Valleys: Chimneys and vertical walls should be properly flashed with metal flashing. All vents should be flashed with prefabricated metal or vinyl flashing. Valleys should be replaced with sheet metal or a double thickness of matching, granule-surfaced, #90 valley shingles.

Costs

The cost of a composition shingle installation varies according to the following conditions:
- the slope of the roof
- the type of deck
- the configuration of the roof (simple or complicated)
- hip or gable roof
- solid or spaced sheathing
- laminated or lightweight shingles
- single story or multi-story building
- tear-off, recover, or new

See Chapter 18 for further information on costs and cost estimating.

Wood Shakes and Shingles

Almost all of the earliest structures in America were roofed with cypress or hickory shakes, hand-split by pioneer builders, usually on the job site. By the end of the nineteenth century, most municipalities banned the use of wood roofs, because of the danger of fire. After the infamous Great Chicago Fire, widespread use of asphalt shingles, slate, and terneplate almost completely eliminated the use of wood shingles, except in restorations of historical structures, when the aesthetic appeal of the split shake roof is desired in new construction.

Physical Properties

Wood shingles are *sawn* wood products, generally of cedar, cut in 16", 18", and 24" lengths, with uniform butt thickness. Wood shakes are split wood products, with a variable butt thickness. Shakes are categorized as #1 Blue Label (premium grade) for roofs and sidewalls, #2 Red Label (good grade), #3 Black Label (utility grade) for economy applications, and #4 Undercoursing, used for starter courses only. Because of the low material cost in relation to the labor cost for shakes, #1 Blue Label is almost universally specified for roofing.

Wood shakes are split from short sections of a log, and shaped by the manufacturer into one of three different types: *hand split and resawn* (the most common), *taper split*, and *straight split*. The first type, split and resawn, is broken into two further classifications: *#1 handsplit and resawn* are made by cutting cedar logs into desired lengths, usually 18". Blanks are split from the logs, and then sawn diagonally to produce two tapered shakes from one blank. The second type, #1 Taper Sawn, are sawn on both sides. Taper split shakes are shaped using a sharp-bladed steel froe (cleaver) and mallet to form a tapered shake. *Straight-split* shakes are shaped by machine or hand in such a manner that the shake is straight like a shingle, and not tapered.

Fire Resistance

Both wood shingles and wood shakes are available in pressure-treated wood to meet U.L. 790 fire-rated standards for Class A roofs. However, this treatment nearly doubles the cost of the product, which limits its use in the marketplace. Also, there is evidence that the fire-retardant chemicals are leached out of the cedar over time, rendering the product less resistant to combustion. Fine-mesh screens are commonly placed over chimneys and stacks to catch any sparks that might ignite this type of roofing. However, many codes prohibit the use of shakes and shingles due to the risk of fire.

Decking

Both solid and spaced sheathing are used as decking for wood shingles and shakes. Spaced sheathing provides superior ventilation of the shakes, thereby helping to prevent rotting. Spacing of the 1" x 4", or 1" x 6" softwood sheathing boards should be the same as the weather exposure of the shingles or shakes. However, at the eaves, 36" of solid sheathing with #30 felt should be provided. A layer of roofing felt is required to prevent wind-driven snow or foreign matter from entering the attic. Omission of the felt baffle (see Figure 12.2) is the most common mistake in the application of wood shakes, and causes many premature roof failures. To be effective, the top of the felt baffle must rest on the sheathing. When using spaced sheathing, nails should be driven through the upper portion of the sheathing with the felt baffle attached to the lower portion.

Installation Considerations

As consumers have expressed a renewed interest in wood roofs, many inexperienced installers are taking up this business. Their lack of skill, together with a failure to follow some basic principles of wood shingle installation, can lead to major problems. For example, architects and designers may not be aware that a minimum of 4" in 12" slope is required to properly drain a wood roof. The inexperienced roofing contractor working with incorrect specifications may compound this error.

Two nails per shingle are required, as are the previously mentioned *baffles*. Nails should always be galvanized or copper, placed one inch from each side and just high enough to be covered by the next course of shakes (see Figure 12.2). Nails should be driven flush to

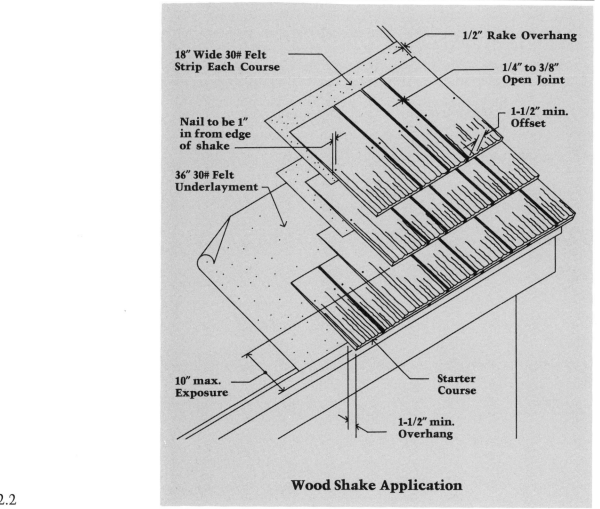

Wood Shake Application

Figure 12.2

the surface and not into the fiber of the wood. The nails should penetrate the wood substrate by at least $^1/_2''$.

Sequence: The roof should be pre-loaded with bundles of wood shakes or shingles the day before application is to begin. Thirty-pound felt should then be applied to the entire deck. The installation should begin at the eave with an overlay of 1-$^1/_2''$. Caulk lines should always be struck to ensure uniformity and alignment. The 18″ strips of 30 pound felt (the baffles) should be placed to overlap the shakes by four inches. Cut shakes or shingles should be saved to use along the valleys. Proper planning should provide orderly coursing. Pre-manufactured hip and ridge units can be obtained from the supplier to ensure uniformity. It is recommended that a $^3/_8''$ to $^1/_2''$ space be left between shakes to prevent buckling. Side lap should be 1-$^1/_2''$ for each succeeding course, as shown in Figure 12.2.

Flashings

Flashings should be made of non-corrosive metal, such as copper or stainless steel, and should conform to the standard details shown in the NRCA manual (see Figure 12.3). Step-flashing at walls and chimneys should be carefully installed by an experienced mechanic. It is difficult for a framing carpenter, without experience, to fashion a proper flashing. Usually, the first leakage occurs at the step flashing.

Attic Ventilation

Inadequate ventilation can cause early roof failures, whether wood or asphalt shingles are used. This is a common installation problem. Vents should be as close to the ridge as possible, or along the ridge itself. An equal opening area should be provided in the ridge and soffits, to allow proper attic ventilation.

Summary

As with all roofs, it is important to engage a competent roofing contractor with documented experience to install wood and asphalt roofs. Any prospective roofing contractor should be asked to provide examples of successful projects recently completed. This prequalification should occur before bidding and is an essential step in planning for all types of roof installation.

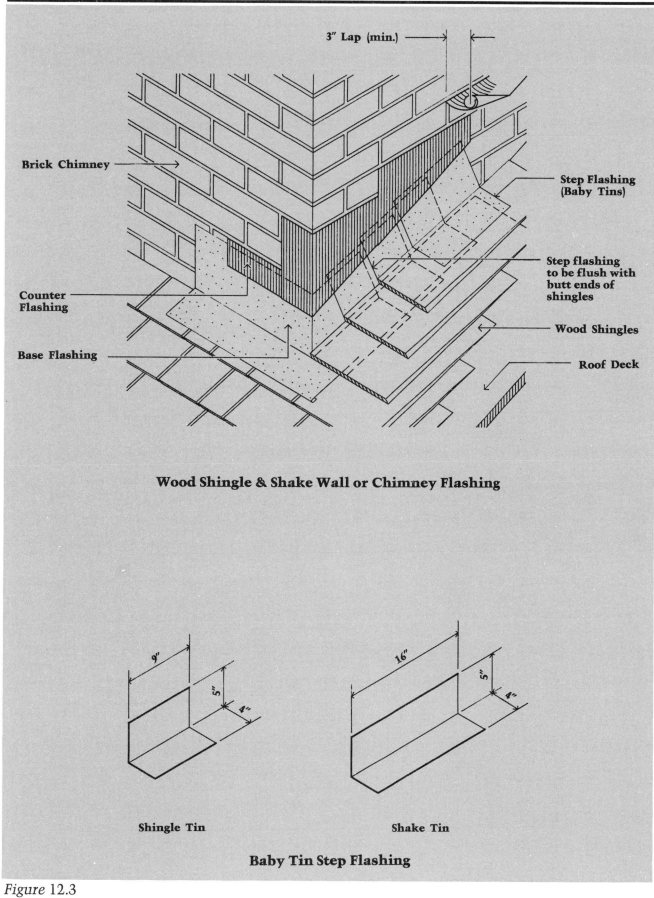

3" Lap (min.)

Brick Chimney

Step Flashing (Baby Tins)

Step flashing to be flush with butt ends of shingles

Counter Flashing

Base Flashing

Wood Shingles

Roof Deck

Wood Shingle & Shake Wall or Chimney Flashing

9"

5"

4"

16"

5"

4"

Shingle Tin

Shake Tin

Baby Tin Step Flashing

Figure 12.3

CHAPTER THIRTEEN
ACCESSORIES

CHAPTER THIRTEEN

ACCESSORIES

Accessories include skylights, curbs, hatches, and smoke vents. They are generally grouped with roofing and insulating materials under the Construction Specifications Institute's Division 7, *Thermal and Moisture Protection* (See Appendix G for a complete listing of CSI Division 7 categories).

Accessories, when properly installed, protect the building interior from moisture, as does the roof membrane itself. When improperly installed, accessories may actually be *sources* of moisture intrusion, by creating discontinuities in the roof membrane.

Ideally, the roofing contractor prepares the roof and flashes the accessories in place after they have been installed by the general or mechanical contractor. The location, exact dimensions, and other pertinent details about rooftop accessories should be discussed at the pre-roofing conference by all parties involved. This information is then recorded in a written document and copies sent to all subcontractors, as well as to the owner's representative, general contractor, and any other affected party.

The final selection of rooftop equipment should be made in the early stages of the project, with coordination between the various trades on the project. This step requires the attention of the project coordinator, whether that person is the project architect or a construction manager. Resolving these details early is important to the smooth and efficient installation of accessories.

Skylights

The use of skylights is increasing in contemporary construction, with the popularity of atriums, naturally-lit common areas, and "sun spaces" in today's office, retail, commercial, and residential environments. As manufacturers sell greater numbers of skylights, they are provided a wider range of flashing and other accessories for the installation of these units. However, many skylight suppliers do not include curbs with their units and, in many cases, this component must be field-fabricated by the general contractor.

Figure 13.1 is an illustration of the components and configuration of a skylight roof curb. It is imperative that curbs be built and located while the deck is being installed. If decks are not cut and curbs not installed until *after* the roof membrane is in place, there is a much greater chance of roofing failure at the point of curb installation.

Other Roof Curbs

Like skylights, other penetrations requiring curbs should also be coordinated early in the roof installation process, preferably during the pre-roofing conference. Items such as smoke vents and hatches should be detailed on construction documents. Without this information, field fabrication and setting of the curbs may result in damage to the roof membrane.

Mechanical and Electrical Pipe Penetrations

The project architect does not always show the locations of mechanical or electrical pipe or conduit penetrations on the architectural drawings. As a result, the plumbing contractor may make roof penetrations close to roof curbs or parapet walls. This positioning makes proper flashing of these penetrations almost impossible. Minimum clearance distances should be established and documented. Figure 13.2 shows the suggested minimum clearance for penetrations provided by the National Roofing Contractors Association.

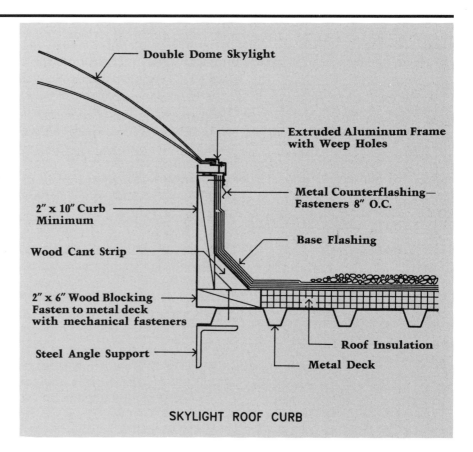

Double Dome Skylight

Extruded Aluminum Frame with Weep Holes

Metal Counterflashing— Fasteners 8" O.C.

2" x 10" Curb Minimum

Base Flashing

Wood Cant Strip

2" x 6" Wood Blocking Fasten to metal deck with mechanical fasteners

Steel Angle Support

Roof Insulation

Metal Deck

SKYLIGHT ROOF CURB

Figure 13.1

As with accessories, the best time to determine the locations of mechanical or electrical roof penetrations is at the pre-roofing conference. There, HVAC, plumbing, electrical, and general contractors can plan the locations of their units. Locations of rooftop electrical switchgear and panels, as well as plumbing vents and refrigerant should be determined prior to installation of the roof.

If possible, the sizes of penetration hoods (see Figure 13.3) required should be noted so that the sheet metal subcontractor may fabricate them in advance. Mechanical penetrations, such as refrigerant lines and control wire conduit, should be grouped with electrical conduit in order to decrease the number of penetration hoods required.

Occasionally, mechanical equipment is placed on bases such as the one shown in Figure 13.4, rather than using a continuous curb or base. These bases should be flashed in the same manner as any pipe penetration, except that nailers and insulation should be used to isolate the supporting base from the deck, as shown in Figure 13.4.

Accessories and Single-ply Roof Seams

With single-ply roof (SPR) systems, it is especially important that the roofing contractor study the roof plan and coordinate the location and sizes of all rooftop accessories. If a curb or other penetration should occur at the seam between two sheets of SPR membrane, the likelihood of a leak is greatly increased. The roofing contractor should indicate on the drawings at the

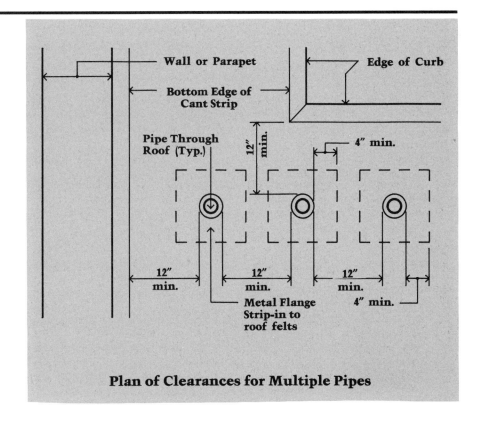

Plan of Clearances for Multiple Pipes

Figure 13.2

Gutters and Downspouts

pre-roofing conference the locations of all SPR seams, if SPR is being used. When locating field seams, allowances should be made for factors such as edge sheets that are to run up parapets.

Depending on traditional practice in the area, gutters and downspouts may be installed by a subcontractor, instead of the roofing contractor. This division of responsibility may complicate some types of projects, in which case it is best for the roofing contractor's own crew of sheet metal workers to perform the gutter installation. In any event, coordination of the roof installation with gutter and downspout placement is critical; gutters should be positioned before the roof edging material is installed.

The best installation procedure entails running a strip of vinyl down into the gutter, under the edging, so that moisture will not find a path behind the gutters. This strip is shown in Figure 13.5 behind the gravel guard. The gutter shown in this figure is an *Ogee Configuration*, which is more rigid than a box gutter. This shape usually can be obtained in 20 or 30-foot lengths. Use of these lengths eliminates many joints, thereby reducing the cost and decreasing the potential for leaks. Material in 26-gauge is adequate for gutters formed out of roll stock up to 15″ wide.

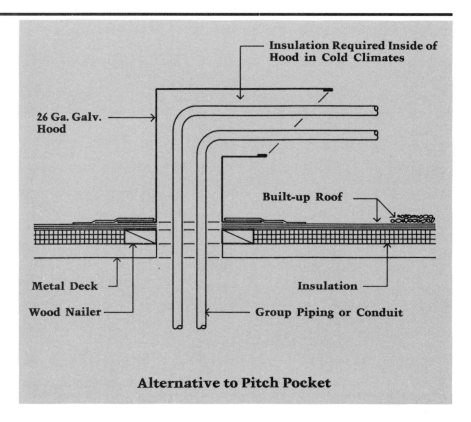

Alternative to Pitch Pocket

Figure 13.3

138

Gutters Versus Scuppers

Because of problems with maintenance, deterioration, and cleaning, it is best to avoid the use of gutters when possible, especially on large buildings where they may be required to move the water a great distance, thereby increasing the chance of blockage by debris or ice. The best way to remove water from the roof is with interior ("field")drains. These drains should be installed along with tapered insulation, to ensure a positive slope and thus minimize ponding. Short lengths of gutter, *scuppers* or *heads*, and downspouts are another solution. With this approach, water is directed away from the raised portion of the fascia by the use of *crickets* between the scuppers (see Figure 13.6). The most economical downspout is the 26-gauge corrugated rectangular

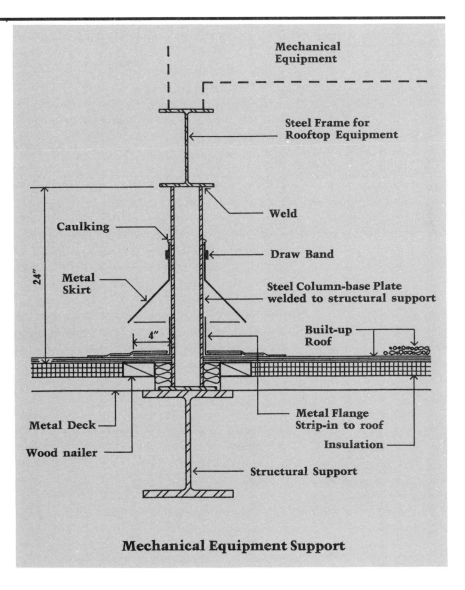

Mechanical Equipment Support

Figure 13.4

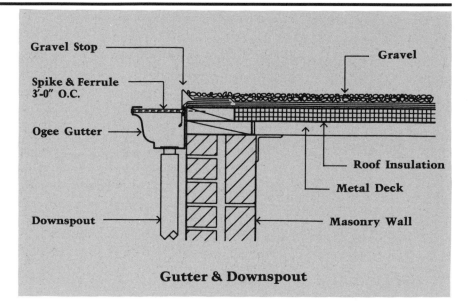

Gutter & Downspout

Figure 13.5

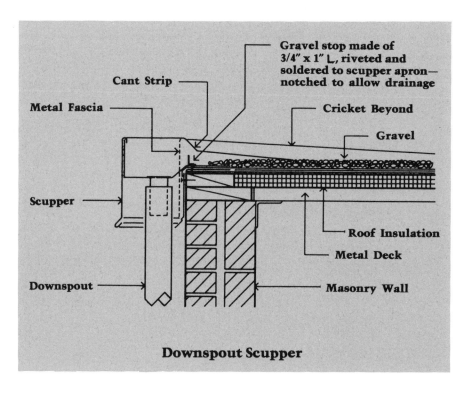

Downspout Scupper

Figure 13.6

140

type, readily available in most areas at sheet metal supply wholesalers. These downspouts are usually available in three sizes:

Area in Square Inches	Actual Size	Nominal Size
7.73	2-$\frac{3}{8}$" x 3-$\frac{1}{4}$"	2" x 3"
11.70	2-$\frac{3}{4}$" x 4-$\frac{1}{4}$"	3" x 4"
18.75	3-$\frac{3}{4}$" x 5"	4" x 5"

Scuppers should be spaced no more than 50 feet on center; they should be closer if there are large areas to drain. The area to be drained by each downspout should be calculated. Then, using rainfall data derived from charts in the *SMACNA Architectural Manual*, *AIA Graphic Standards*, or a similar resource, the size of the downspouts can be determined, and the distance between the downspouts adjusted as necessary to produce practical spacing. If the downspouts are over 40 feet in length, relief valves or heads should be installed to admit air and prevent formation of a vacuum.

Gutter Materials

The choice of gutter material is based on both cost considerations and design factors. The most cost-effective material is 24 or 26-gauge galvanized steel, which can be easily worked and soldered. Heavier gauges, such as 20 or 22-gauge, may appear more sturdy, and hence, worth the extra cost. However, heavier gauges are difficult to break properly. They expand and contract at greater rates, and they are cumbersome to work with. Using stainless steel or copper may avoid the necessity and cost of painting, but these are expensive choices, and copper will stain the face of the building with copper sulphate if runoff occurs. Aluminum should be avoided since it cannot be soldered and has a high coefficient of expansion and contraction.

Summary All details that affect the integrity of the roof membrane should be discussed in the pre-roofing conference. Detailing of mechanical and electrical penetrations, as well as drainage structures, should be reviewed by subcontractors. This planning session should result in an orderly flow of sheet metal work placement of all curbs and rooftop structures or penetrations prior to installation of the roofing. Architects or project managers are well rewarded for their efforts regarding the conference and follow-up by greater cooperation between trades, a more weathertight structure, and a more accurate set of as-built drawings for the building owner.

CHAPTER FOURTEEN

MAINTENANCE AND RESTORATION

CHAPTER FOURTEEN

MAINTENANCE AND RESTORATION

The Importance of Maintenance

Most building owners tend to ignore the building roof until a failure occurs. The notion that their manufacturer's warranty will take care of any problem is ill-advised. The roof warranty is normally a carefully worded document that puts a great amount of responsibility on the installing contractor for the successful performance of the product. The roof is normally guaranteed against "leaks" for a period of ten or more years, providing "reasonable" roof maintenance is executed by the owner during the term of the warranty. This type of clause, although acceptable, is subject to a broad interpretation and should be clearly understood by the owner. Some manufacturers require their installers to affix a sign at the roof access explaining these conditions.

Damage to the roof by tradesmen, or "acts of God" are not covered by warranties. Oil or chemical spills, punctures, vegetation damage, structural movement, even ponding can each be sufficient reason for the manufacturer to void the roofing warranty. If a roof should leak, causing damage to or destroying valuable building contents, the roof warranty does not pay for replacement of these contents — only for restoration of the roof itself. If a roof is badly blistered, with felts exposed to degradation as a direct result of poor installation, the warranty is also void because it covers failures attributable to the materials, and not to installation defects or oversight. It is, therefore, imperative that blisters or other early-stage problems in built-up roofs be repaired as soon as they are detected.

Roof Depreciation

Since the roof is probably the largest single depreciating element in the average building, and is expected to require replacement at least once within the life of the building, it makes sense to have a comprehensive maintenance and restoration program. The new tax laws in effect at this writing require investments of this type to be depreciated over a period of 31 years, rather than the 19 years formerly allowed. This change should have a significant impact on future operating policy for building managers. If it is prudent to have a maintenance contract for equipment valued at a few thousand dollars, it seems more prudent to have a maintenance program for the roof system, the replacement of which is often valued at hundreds of thousands of dollars. The increase of preventive maintenance contracts supplied by roofing contractors is a clear indicator that owners are acknowledging the importance of ongoing roof maintenance.

There are essentially two types of maintenance contracts available today; these are *service contracts* and *maintenance agreements.*

Maintenance and Service Contracts

Service Contracts

The simplest form of contract is the *service contract*. In this arrangement, the roofing contractor offers a price, usually by the hour, to the owner, along with a pledge to promptly respond to the owner's call and make repairs to the roof as necessary. No guarantee is involved other than the promise that the work will be done properly and by skilled mechanics at a prearranged cost.

Maintenance Agreements

The second approach is the *maintenance agreement*. In this type of agreement, the roofing contractor agrees to keep the roof "free of leaks" for a specified period of years. At the outset, the maintenance agency makes a comprehensive inspection of the roof using moisture detection devices, visual checks, and all other diagnostic methods available. After the history of the system has been reviewed, a cost is established to enter into the agreement. If necessary, the roof is first repaired where faults already exist that may necessitate early action. The maintenance agency determines the warranty charge per square foot, or as a lump sum. A percentage of this fee is normally paid upon execution of the agreement, and the balance is paid over the term of the agreement.

The building manager or owner should avoid "front end" payment of an entire maintenance agreement, even if the roofing company is reputable, unless this agreement is backed by a proven roof system manufacturer. In roofing, as in other areas of construction, the company that least deserves the trust of the buyer is often the one that is in greatest need of cash flow. If a little more is paid at the outset to enlist the services of a knowledgeable, experienced roofing firm —which may not be the low bidder —continuing product support is not in question. In any case, semi-annual inspection (as well as inspections after any significant weather event) should be part of any effective roof system maintenance program.

The Owner's Roof File

The best roof maintenance program is one that is initiated and controlled by in-house personnel. The facilities manager or maintenance supervisor should compile a "roof file" for each building, or group of buildings, in a roof installation. This file, if properly documented and updated, will become as valuable a document to the owner as the warranty.

File Contents

As a bare minimum, the Roof File should contain:
- Roof Plan, showing the location of all penetrations, equipment, etc.
- Location of all seams, in single-ply systems.
- Construction specifications (as-built) and related documentation.
- Copy of the pre-roofing conference proceedings, agreements and any addenda thereto.

- Copies of all approved submittals and brochures on each component of the system.
- Roofing system bonds or guarantees, whether from the manufacturer or contractor.

Initial Warranty Period

Normally, the major manufacturers require that a warranty be effected and issued, even if the owner either does not require it, or intends to use in-house personnel for maintenance and repair. Most warranties require that the installing roofing contractor be responsible for the roof for the first two years of a warranty period. Any deficiencies should be recognized within this time and corrected by the original roofing contractor.

Roof Inspections

After the initial warranty period (normally two years), a manufacturer's representative inspects the installation. This representative should be accompanied by the building supervisor or facility manager. During this inspection and at all subsequent annual inspections, a checklist should be completed, dated, and included in the File. The roof, as a whole, should be checked for debris, discarded tools or parts, construction materials, broken glass, limbs, leaves, nests, or other material that may damage the roof or block roof drains or gutters.

Structural or Pending Problems
During annual inspections, structural change can be observed. Noting any movement of walls or settling of the deck is particularly important. Cracks in walls or splits in the roof membrane might indicate differential movement in the structure itself. Excessive ponding would suggest deflection in the deck, and should always be corrected. The measures taken may include structural modifications to correct the deflection, or installation of tapered insulation board. "Expansion joints," or *area dividers*, might be the solution for splitting. Early attention to these problems could prevent premature replacement of the roof. The best solution by far, however, is to avoid problems through careful design and selection of structural and roof system components.

Repair of Blisters in BUR
The roof should also be checked for telltale signs of problems, such as blisters or *mole runs*. As discussed in Chapter 2, blisters and moleruns normally indicate a wet substrate. Tearing and displacement of the felts and flashing might suggest poor attachment of the insulation to the deck. Separation of the felt plies in a BUR might suggest that asphalt at too-low a melting point was used for the slope of the roof. Early detection of these problems can allow solutions that may not be feasible at a later date.

Felts are often left exposed after a repair, subjecting them to additional degradation. Blisters that occur should be cut in an "x" and the "flaps" folded back, to allow as much drying as possible. This cavity should then be filled with compatible materials and the flaps pressed back into place. Two plies of polyester felt membrane should then be mopped in over the fissure. It is probably not cost-effective to repair small blisters (under one square foot area). Modified bitumen membrane material containing polyester mat reinforcement is ideal for repairing such blisters and splits, since it is tough and can be torched into place.

These repairs are not normally covered by the roof warranty, but to disregard them jeopardizes the service life of the roof.

Flashings Versus Seams

Annual inspections should pay special attention to flashings and seams in SPR systems. The experienced inspector normally focuses on seams because history has shown that this is where most problems occur. The flashings for single-ply roofs should be checked for separation, bridging, excessive stretching, and oxidation.

Neoprene (uncured) flashing was first used by most EPDM manufacturers. This material often failed in its early use due to ultraviolet degradation. Consequently, most flashings for single-ply roofs are now uncured EPDM, which offers better chemical compatibility, resistance to ultraviolet light, and a longer service life. Flashings on built-up roofs should be checked for delamination, slippage, loss of surface protection, stretching, and in-service damage, such as heel marks, fractures, or splits that might allow moisture into the substrate.

Expansion and Contraction Problems

Latent design-related problems may exist when the wall is independent of the roof structure with no isolation provision for expansion and contraction at the wall. Pitch pockets and pourable sealer pockets should always be checked for adequate seal.

Surface Contaminants

Oil spills or other contaminants on the roof, such as exhaust exposure, or precipitated grease from kitchen exhaust hoods, should be monitored for adverse effects on the membrane, and corrected if necessary. Preventive maintenance of this sort can arrest problems before severe and irreversible damage occurs to the roof.

Subsurface Contamination

If, from inspection, it is established that a roof has been leaking for some time, the insulation and substrate should be checked for moisture contamination. This can be done by using either a two-inch diameter core sampler, or a non-destructive moisture tester — either nuclear, infrared, or capacitance type — which indicates the moisture content of the substrate. The core sampler removes a cylindrical core of the roof membrane and insulation. The sample is weighed wet, then dried in an oven and weighed again. The quantity of moisture per cubic foot of insulation can thus be calculated. At the NBS-NRCA symposium in April, 1979, a method of correlating moisture content with thermal resistance was presented. At 100 percent moisture content by weight, a wet sample may weigh approximately twice what it weighs dry.

Most common insulations, such as perlite board, fiber board, and glass fiber board may lose over 50 percent of their insulating value when wet. Moisture eventually breaks down bonding agents in the insulation, causing it to collapse. This loss of dimensional stability of the insulation may cause it to tear the roof membrane. Wet insulation should be removed and replaced, with subsequent repair or replacement of the roof membrane.

Venting as Means of Drying: It was once generally accepted that moisture vents (or solar activated relief vents) were capable of "drying out" wet insulation. However, Wayne Tobiasson of the Cold Regions Research and Engineering Laboratory has determined (based on experiments) that edge and membrane (top surface) venting offers no significant benefits for removing moisture. Vents in a roof over wet insulation allow only a small area around the vent to dry; they have not been proven to significantly dry out more distant insulation. Significant drying of roof insulation can only occur from below, if there is no vapor retarder, or the deck itself is permeable.

Resaturating and Coating

The owner on a tight budget has several repair options with regard to built-up roofs, both smooth and gravel-surfaced. To avoid replacement of the roof, which can be costly, the owner may consider recoating or *resaturating* the roof. Resaturating a roof includes the removal of all gravel from the surface, followed by the application of up to seven gallons per square of cut-back asphalt or tar pitch coating that purportedly penetrates the top plies of the roof and restores their waterproofing qualities. Fresh gravel is then embedded in this coating. Clearly, if the existing roof is badly split or blistered, resaturation might tend to simply forestall the problem rather than correct it. Since resaturation can cost as much as 50 percent of the cost of re-roofing, it is wise to seriously consider re-roofing whenever resaturation is a possibility.

Asphalt coatings have been developed for use in the restoration of smooth-surfaced built-up roofs. Asphalt coatings may contain mineral fibers to add durability, or aluminum powder to produce a heat-reflective surface. These coatings are relatively inexpensive and can add years of useful life to the roof if applied before moisture has penetrated the felts. Some roofing contractors may talk an owner into coating an "alligatored"roof with hot asphalt. However, hot asphalt will not penetrate cut-back coatings, and has a negative effect as it adds stress to the membrane.

Recently, roof paints and elastomeric coatings have been introduced as a panacea for all roofing problems. These coatings are being used primarily to seal leaking metal roofs, and at fasteners and seams, with a 30 to 50 mil. thickness of elastic materials. It is impossible, however, to control the thickness of such coatings where it is most needed, on ribs and seams.

Roof Restoration

Conversion of historic buildings demands special expertise and attention to detail, along with the use of ancient techniques and skills. This work not only requires special techniques, but it is labor-intensive as well. Built-in (Queen Anne) gutters, decorative soffits, cupolas, finials, eyebrows, lentils, gargoyles, and other common architectural adornments of the nineteenth century are back in demand. Restoration or matching of standing seam roofs of copper, terne metal, or stainless steel require the skill of experienced mechanics.

Tile Roofs

Tile roofs present their own particular challenge to roofing contractors. Tile was often laid over interlocking wood lath and not nailed to the wood substrate. After 50 years and more service, the underlayment beneath tile roofs, which serves as the waterproofing course, may be deteriorated. This requires that each tile be carefully removed, marked, and saved for reinstallation after the underlayment is restored.

Most interlocking tile roofs are no longer commercially available and require special hand-made molds and special firing to produce. Substituting available tile to approximate the appearance of the original tile requires imagination and skill of the roofing contractor.

Structural Work

Often the decking and supports, such as beams and joists, are rotten and require replacement in the course of roof restoration. Many roofing contractors shy away from structural work, feeling that they do not possess the requisite skills. Specialty remodeling contractors, who may have less roofing skill, but more nerve, may perform this kind of work. Those roofing contractors who are willing to take a chance on restoration and all that it entails may get "choked" a few times, but if tenacious, may find the work not only profitable, but very rewarding.

Metal Shingles, Cornices, and Fixtures

Replicas of stamped metal shingles, cornices, and ornamental metal fixtures tend to be difficult or impossible to locate and may require custom fabrication. Consequently, a roof restoration contractor should have available a fully equipped sheet metal shop. Low cost is seldom a major consideration in restoration work; however, labor-saving equipment, such as seamers and pan formers (roll formers), should still be used.

The First Phase of the Restoration

In restoration work, the roof is commonly brought to a watertight condition before any interior work is performed. An architect usually prepares a set of plans and specifications outlining the work to be done. This first phase of the work, including any deck repairs and painting, is bid separately from the other renovation activities. Since the roofing contractor has the largest share of this work, it is a rare opportunity to bid as the prime contractor. Acting as prime contractor requires a knowledge of the paperwork and bonding and insurance requirements. The roofing contractor who has mastered these aspects of contracting can build a reputation in a specialized and growing field.

Unfortunately, most restoration work is loaded with unseen pitfalls for those who are inexperienced. The buildings involved have an intrinsic value and often contain priceless objects. Therefore, the integrity of the roof must be preserved.

The Need for Public Relations Skills

The successful pursuit of restoration work requires excellent public relations skills. The agencies responsible for historic buildings often wish to be involved in every detail. Therefore, it is vital that the roofing contractor establish a relationship with one individual, preferably an architect, to monitor the exclusive contract on the job. The best time to do this is at the mandatory pre-bid conference.

Scheduling and Safety Concerns

Scheduling: Time constraints imposed on the roofing contractor can be another problem with restoration work. Careful restoration is a time-consuming business. Often, the materials required can take months to be delivered. The prudent roofing contractor will carefully consider all such factors before undertaking the work. One problem may be owner and/or architect-imposed liquidation damages for delays in completion of the work. Such conflicts can be avoided if a reasonable completion schedule is agreed upon before the work is begun.

Experience with restoration work is the key to the preparation of an accurate schedule. Many factors never dreamed of by the estimator can delay or even stop progress of this work.

Safety: Restoration not only requires special attention to detail, but it often involves working high up on steep roofs. Parts of the job may have to be done from scaffolds or hanging from a boson's chair attached to a crane. Safety belts and lifelines are required equipment. Towers and steeples are often infested with wasps which can prove dangerous to the mechanics working high on scaffolds. Sudden gusts of wind can also be very dangerous to mechanics perched high on steeples. Dry rot and decay can cause entire sections of decking to cave in when the old roofing is being removed. Boom hoist equipment and other similar devices must be used in the effort to avoid costly overruns, injuries, and potential catastrophe on the job site.

Many restoration projects require the removal of existing roofs, some of which may contain asbestos fibers. The asbestos problem cannot be ignored, since the consequences to both contractor and owner can be devastating if not dealt with early in the planning stage. If this material is found in any form, the proper agency must be notified, samples taken and analyzed, and the proper removal and disposal procedures followed. The regulatory requirements and safeguards are constantly being updated or revised; regulations or requirements in one part of the country may not be the same as in another area. The National Roofing Contractors Association (NRCA) continues to monitor the asbestos situation and provides regulatory information as it becomes available. See Appendix A for more information on NRCA.

Other Considerations

Any project over 50 years old may have historical significance. Roof restoration projects may range from a water treatment plant to a train station. The approach to the work should be organized and cautious, and the roofing mechanics chosen for the job should have the skills to perform the work in the amount of time allotted. Poor performance can damage the contractor's reputation and eliminate the firm from consideration for future work. Most restoration work is assigned on the basis of referrals, not price.

Prequalification of the restoration contractor can be achieved by placing a requirement in the specifications that the contractor show evidence of recent completion of similar projects, including references. Skill, financial strength, and experience should be as important as the price in the bidding process.

Summary

There is no question that roof restoration is difficult, and requires skill, commitment, and experience. Nevertheless, if it is approached cautiously and with intelligent preparation, this type of roofing can prove very rewarding — both financially and in terms of the satisfaction of achievement.

CHAPTER FIFTEEN

WARRANTIES

CHAPTER FIFTEEN
WARRANTIES

Roof Bonds

Enabling Legislation

Roofing warranties issued today are regulated by the *Magnuson-Moss Act* of 1975, enacted by Congress to control consumer product warranties. Before the advent of today's *full value warranties*, the industry normally used what was termed a *bond*. The bond guaranteed that the roof would be maintained in a "watertight condition" for a specified length of time (usually 20 years), subject to certain conditions and limited by a maximum liability, called the *penal sum*.

Manufacturer and Surety Company Liability

Before the Magnuson-Moss Act, the liability in case of a failure of the roofing system was assumed by the surety company and not the manufacturer. However, it was the responsibility of the manufacturer and its agents to establish a network of approved roofing contractors and to assure the quality of the installation. In practice, construction of the roof system was rarely monitored by an agent of the manufacturer, whose primary focus remained sales. Manufacturers soon learned that inferior roof installations resulted in increased failures, which, in turn, eroded sales. Consequently, manufacturers became more selective in contractor recruitment.

"Penal Sum" Versus Full Replacement Value

When a roof failed due to "ordinary wear" or "exposure to the elements" during the term of the bond, it was the owner's responsibility to notify the manufacturer in writing. The manufacturer would have the roof inspected and repaired by an approved contractor. Generally, the cost of most repairs was relatively small, but if an inspection indicated a need for the complete replacement of the roof, the surety company was informed and issued a check in the amount of the full penal sum. In times of high inflation, this sum frequently amounted to only a fraction of the actual cost of replacement; the owner frequently was required to pay the sometimes substantial difference. Many times, this came as a surprise to owners, who normally assumed the roof was bonded for 20 years against additional expense.

"Implied Warranty" Litigation

Although owners seldom realized it, roofing contractors frequently were guilty of offering "implied warranties" when an installation was represented as a "20-year installation." As test cases exposed these practices, contractors became careful in the use of the term "20-year roof."

When manufacturers began to experiment with coated felts, the incidence of "implied warranties" increased. These built-up systems exhibited increased frequency of splits and blisters, usually occurring within the first five years after installation. Numerous court settlements requiring manufacturers to bear all legal costs and the difference between the penal sum and the total value of the implied warranty, were a major factor in the revolution in roofing systems and the development of single-ply technology in the mid-1970's.

Common Exclusions

The controlling terms of roofing bonds normally appear in the fine print of the warranty instrument. *Consequential damages* (roof leak damage to building contents and portions of the building other than the roof) are excluded. If the use of a building is changed, then the bond becomes void. Most bonds are nontransferable, which means that the bond is cancelled when the building ownership changes. Many common alterations or actions on the part of building owners or managers have the effect of voiding the warranty. Examples of such changes include:

- Traffic pads, promenade decks, or recreational areas placed on the roof after the bond is issued
- Roof repairs or additions made without the consent of the manufacturer
- Repairs or alterations made by a roofing contractor not approved by the manufacturer
- Movement or failure of the substrate
- Failure of any material not supplied by the manufacturer
- Infiltration of moisture into the substrate through walls

Many of these exclusions, common to "old form" 20-year bonds, are still incorporated into roof warranties today. Specifiers and owners should read the fine print and analyze the implications of all such exclusions.

Guarantees and Warranties

The terms *guarantee* and *warranty* are used interchangeably in the industry; there is no legal distinction between them. The term *warranty* is most often used today. *Full value warranties* cover both material and workmanship; *material warranties* or *limited warranties* cover some or all material used, only. In limited warranties, the manufacturer takes no responsibility for the application of the materials — and is responsible only for failure of the *product itself* to keep the building watertight (prevent leaks) for a specified period of time.

Proration

The manufacturer's liability decreases during the warranty period, or is *prorated*, depending on how much of the total term of the warranty has expired. The manufacturer assumes a proportional percentage of the materials replacement cost, if the manufacturer's inspection has proven that the roof was properly installed. It is the manufacturer's prerogative to determine if the installation meets their standards.

Problems Not Covered

Implied warranties as to the fitness of a material for a particular purpose are specifically denied. As stated earlier, consequential damages are normally excluded as well. When researching exclusions in warranties, the specifier will find that most manufacturers do not warrant against discoloration of their product caused by environmental conditions such as dirt, pollution, or biological agents. Loss of granules, delamination, and other similar problems are not covered by the warranty, unless the roof is leaking. Metal flashing, such as gravel stops, are usually excluded from the warranty, unless furnished by the manufacturer. This can lead to problems when leaks develop at the roof edge. One manufacturer supplies hard rubber edging which is covered by the warranty. This approach eliminates the problematic division of responsibility at the roof perimeter. For 15-year warranties, most manufacturers require that all components of their system be used, including fasteners, insulation, and flashing. Some systems specifically exclude applications in which exposure to oil contamination or high winds is likely. Heavy roof traffic might also rule out some systems.

All warranties cover only material furnished by the manufacturer. Some manufacturers refuse to warrant their roofs when installed over another manufacturer's roof insulation.

Reliability of the Warranty

A new manufacturer may not have the financial reserves to cover the obligations assumed in warranting its new products. This is the most important factor in determining the value of a warranty. If the manufacturer is a relative newcomer, or undercapitalized, the warranty has little worth to the building owner. There are some "full value" warranties offered today which are issued in the name of "shell" corporations, and have no value whatsoever. The specifier should investigate the financial condition of the corporation issuing the warranty.

Manufacturer Field Support

Recognizing the need for quality control in installation, but unwilling to remove warranty limitations, most major single-ply manufacturers require inspection of all installations by their own trained agents. They also require training of all installers, both at the factory and in the field. Factory incentives for installers and dealers exercising high quality standards are frequently offered. This quality control and factory support, achievable due to the homogeneous nature of single-ply membrane, has led to tremendous growth in this sector of the roofing market.

Manufacturers' Product and Specification Changes

Past problems with warranties have fostered technology changes in the industry. For example, one major manufacturer of EPDM systems, recognizing that the seams are the most important element in the construction of the single-ply roof, has added a requirement for an additional sealant in the seam to protect the contact seam adhesive from moisture invasion. There has also been a shift in preferred flashing material, from neoprene (which is susceptible to ultraviolet light) to the more resistant EPDM. New requirements, such as that calling for a separation sheet between "hard" decks and the EPDM membrane, recognize that surface irregularities in concrete (or other hard surfaces) could penetrate the elastomeric sheet.

Manufacturers are continually searching for ways to improve their systems and increase dependability. Specification changes are published periodically for use by installing contractors. The specifier should request a list of approved contractors from the selected system manufacturer and allow only those contractors to bid the work.

The National Roofing Contractors Association (NRCA) publishes the *Roofing Materials Guide*. This publication presents and updates information on roof membrane warranties. There are currently over 65 manufacturers and over 180 systems represented in this manual. After determining the financial stability of the manufacturer, the specifier should consult this guide for specific coverage exclusions in each proprietary warranty.

Summary

A roof warranty is only as reliable as the manufacturer and surety company issuing it. No warranty can substitute for competent and careful workmanship in installation, and regular maintenance and inspections. One of the most important responsibilities of the specifier is that of qualifying installers and systems based on their financial stability and commitment to quality.

CHAPTER SIXTEEN

CODES AND STANDARDS

CHAPTER SIXTEEN

CODES AND STANDARDS

Resources

In order to stay abreast of changing standards and ordinances and to properly communicate with one another, roofing designers, specifiers, consultants, inspectors, and installers should have the following resources at their disposal.

- The latest edition of the applicable Building Code(s)
- The Factory Mutual System Approval Guide
- The U.L. Building Materials Directory
- The U.L. Fire Resistance Directory
- The ASTM Book of Standards, Vol. 04.04, *Roofing, Waterproofing, and Bituminous Materials*
- The NRCA Roofing & Waterproofing Manual
  (See Appendix A for addresses and telephone numbers of agencies.)

Importance of Standards

The official standards can significantly affect the materials and methods used in new roofing. Contractors who are not familiar with the latest industry methods will not be able to accurately bid a roofing project or comply with up-to-date specifications. In many cases, installers ignore code references and assume that the system specified will meet the code. Some designers and specifiers, ignorant of codes and standards, make frequent reference to them in their specifications, but call for materials and application methods that will not meet codes and standards. It is vital to understand the scope of these standards and their enforcement before either putting them in a specification or trying to interpret them when they appear, sometimes gratuitously, in specification "boilerplate."

Importance of Code Compliance

A typical regional building code is the SBC, published by the Southern Building Code Congress International, Inc. in Birmingham, Alabama, and used by many states in the southeast. Its stated purpose is the same as that of all the codes: to protect the public's life, health, and welfare in the built environment. Many of the requirements in the codes have an important, although not always obvious, economic impact on the roofing contractor. For example, most (but not *all*) roofing contractors are aware that a "thermal barrier" should be placed between a steel deck and expanded polystyrene insulation. This requirement appears in the SBC (page 120, paragraph 717.1.3): "Foam plastic, except where otherwise provided, shall be separated from the interior of a building by an approved thermal barrier of $^1/_2$ inch gypsum wallboard..." Standards such as this apply to all construction, whether specified or not. Most municipalities do not have a large enough inspection staff to enforce all code regulations, so code violations may go uncorrected on occasion. However, if a building disaster occurs that can be traced to noncompliance to codes or negligence on the part of the designer or installer, these parties may be liable for large judgments.

Factory Mutual

The *Factory Mutual Research Corporation (FM)* was instituted for the purpose of examining construction materials to determine whether or not these materials meet certain performance criteria. Systems are subjected to tests and if they comply with preset standards, these materials receive the FM stamp of approval. For instance, beginning on page 412 of the 1984 *Approval Guide*, the approved roof coverings are listed. The first listing is for *Class I Fire & I-60 Windstorm-Rated (Minimum)*. These coverings are for use over four categories of substrate: 1) structural or insulating concrete, 2) existing roof cover, 3) insulation on steel deck, and 4) insulation on FM-approved plywood or solid lumber deck.

Under the listing for each material, the acceptable insulations with the number of fasteners required are also listed. Acceptable fastener types are designated, but the pattern of fastener is not. Insulation manufacturers normally illustrate the fastener pattern required to achieve either I-60 or I-90 FM standards.

Factory Mutual publishes the following Construction Bulletins relating to wind uplift forces and fire resistance properties of roofing materials:
- FM Bulletin 1-5 Steel Roof Deck Sample for Calorimeter Evaluation
- FM Bulletin 1-7 Wind Forces on Buildings
- FM Bulletin 1-28 Insulated Steel Decks
- FM Bulletin 1-29 Single-Ply Systems
- FM Bulletin 1-47 Roof Coverings
- FM Bulletin 1-49 Perimeter Flashing
- FM Bulletin 1-48 Repair of Wood and Cementitious Roof
- FM Bulletin 1-50 Zip-Rib Roofing and Siding
- FM Bulletin 1-52 Field Uplift Tests
- FM Bulletin 1-54 Roof Collapse

If the designer specifies a generic single-ply system and insulation board over a steel deck, and requires that the system meet an FM I-60 or I-90 approval standard, the installer should refer to the latest *FM Approvals Guide* and the manufacturer's published fastener spacing chart for I-60 or I-90. This information can be used to check the manufacturer's system, determining whether or not it meets the approval standard for its roofing membrane and/or insulation and fastening system. The Approval Guide may be obtained from the FM Training Resource Center, Order Processing Section. (See Appendix A for complete address).

Underwriters' Laboratory

A comprehensive building materials standards agency, Underwriters' Laboratory (UL) uses its own classifications, distinct from those of FM. UL classifications may be found in their *Building Materials Directory*. While Factory Mutual designates their acceptance classifications in "Class I," the UL uses "A," "B," and "C" classifications. Usually the specifier designates a UL classification "A" as the only acceptable system. These systems are listed in the *Building Materials Directory* under "Roof Covering Materials." For each manufacturer, the approved systems, along with acceptable deck types and roof inclines are listed —for both built-up systems and single-ply systems. Approved materials are stamped with the UL label so that there is no confusion as to their approval at the project site. The *UL Building Materials Directory* can be obtained by writing to Underwriters' Laboratories Inc., Publications Stock. (See Appendix A for complete address.)

American Society for Testing and Materials

It is common practice for architects to specify an *ASTM number* in order to designate a material's required performance characteristics. Most manufacturers of roofing materials list ASTM numbers in their catalogs and stamp them on their materials.

The American Society for Testing and Materials was founded in 1898 to develop standards for all types of materials. Each manufacturer has the ASTM labs test their materials to be sure that they meet ASTM's standards. Manufacturers then include this approval designation in their published specification. To obtain the ASTM publications, one must join the organization. (See Appendix A for complete address.)

As an example of ASTM's influence in the roofing industry, most built-up roof specifications call for a *Type IV* glass fiber felt. This designation refers to ASTM standard D-2178; Type IV is a ply sheet with a breaking strength specified in D-2178 as 44 pounds per inch in both directions.

Asbestos Removal in Roofing

Asbestos, a Greek word meaning "inextinguishable," is a fibrous material formerly used in roofing felts, tiles, shingles, flashings, coatings, and mastics. In 1986, the U.S. Senate passed legislation mandating that the EPA, by October 1987, establish "rules and guidelines" for school districts to "assess, abate, and control" materials containing asbestos in school buildings housing grades kindergarten thru 12. The Asbestos Hazard Emergency Response Act (AHERA) has led to far-reaching efforts to regulate the use and disposal of asbestos.

For decades, asbestos was used in many forms of insulation, building board, liners, and roofing material. Today, through pending legislation in Congress and a multitude of local and state ordinances, agencies (both government and private) seek to control the release of asbestos into the environment. Establishing and carrying out standards for asbestos removal promises to create both confusion and opportunity for roofing and other contractors. A roofing contractor who intends to remove a roof that is composed, even in part, of asbestos, should be aware of the following:

- The Occupational Safety and Health Administration (OSHA) requires that the air quality be monitored on any project where asbestos is handled.
- The EPA has set regulations governing control of asbestos dust.
- Most states have licensing requirements for handling asbestos.
- Most insurance policies exclude asbestos-related claims.
- Most states have requirements for disposal of asbestos; contract

documents normally require the roofing contractor to be responsible for disposal.

- The public is acutely aware of the hazards of asbestos, and adjacent residents may react to asbestos removal activities by filing "cease and desist" actions or other litigation against the contractor.
- Technically, the building *owner* possesses the asbestos, and has the responsibility for its proper removal and disposition.

Agencies That Monitor Asbestos Removal

The regulations for asbestos removal are published by the following agencies: The Occupational Safety and Health Administration (OSHA), which sets exposure limits for workers on the job; the Environmental Protection Agency (EPA), which regulates the use and disposal of asbestos; and state or local agencies. The owner and roofing contractor must become familiar with the applicable state and/or local laws and ordinances for that jurisdiction in which the asbestos-related work is to be performed. Numerous states, including California, Illinois, New Jersey, New York, Ohio, and others have already enacted statutes requiring licensing and/or certification of contractors performing asbestos work. It is likely that similar legislation will be enacted in most states as asbestos-related litigation proliferates.

The NRCA and the National Roofing Legal Resource Center are involved in an ongoing project examining asbestos-related regulations; see Appendix A for more information on these organizations.

Most local regulatory agencies are still not certain of their roles in the administration of asbestos-related regulation. It is often difficult for local inspectors to distinguish "friable"(in danger of crumbling) asbestos, from asbestos that is not friable.

Many states that do not have licensing laws have set up guidelines for asbestos removal. The state of Virginia, for example, has the following guidelines as of July, 1988:

1. NIOSH/MESA-Approved respiratory protection must be worn at all times.
2. Full body coverings (i.e., disposable coveralls) must be worn at all times.
3. Loose debris from the roof must be removed using a wet vacuum system and water, or other dustless methods.
4. Area to be removed is to be sprayed continuously with "amended water" to control dust while removal is under way.
5. All roofing debris must be collected and bagged or wrapped in six mil minimum thickness polyethylene, using dustless methods.
6. All wrapped debris must be *lowered* to the ground, not dropped.
7. Polyethylene sheeting, six mils thick, large enough to cover the ground in the loading area, should be used to store bags/pallets of roofing waste until the waste can be disposed of properly.
8. Air monitoring must be conducted by the contractor pursuant to the requirements of 29 CFR 196.58.

9. The Agency must procure the services of a consultant to monitor the project for the Agency to:
   a. Conduct air monitoring to insure no contaminants are being released due to poor work practice; and
   b. To ensure all sections of the removal specifications are being followed.
10. All applicable EPA and OSHA regulations are incorporated into the Virginia guidelines, including but not limited to 40 CFR, parts 53 to 80; 29 CFR 196.58; 29 CFR 1910,1001, and 29 CFR 1910.134.
11. EPA-approved training and state licensing is required for all those involved in the removal process.

Summary

Researching and complying with the wide variety of applicable codes for roof installation and repair presents a challenge to every roofing contractor. For some types of work, such as asbestos removal, not only are the regulations complicated, but the cost is high, and a careful, organized approach is essential. All roofing contractors, designers, specifiers, consultants, and inspectors should have the latest editions of all applicable codes and standards. This information provides a reliable set of guidelines which, followed carefully, ensures the integrity of this aspect of their work.

CHAPTER SEVENTEEN

ROOF SYSTEM SELECTION

CHAPTER SEVENTEEN

ROOF SYSTEM SELECTION

A Systematic Procedure

The selection of a roof system for a particular building is one of the most important decisions an architect or specifier will make on a project. Today, there are so many new roof systems appearing on the market that a specifier may be inclined to choose the system that is most visible in the latest trade magazines. The best approach is an objective and deliberate determination.

This chapter outlines the roof system selection process and explains how that process affects other aspects of design. *A Roof System Selection Chart* is presented in this chapter and provides general guidelines on the strengths and weaknesses of each of the major systems. Also introduced is the concept of the *Roofing Questionnaire* (See Appendix B), which the architect or building manager may use early in the selection or design process to narrow the field of possibilities.

The Goal: An Objective Selection

As in any design situation, roof system selection is a subjective process. Inevitably, one's personal experience plays a large role in the selections one makes, especially when working with an important client, or when one's job security may be in jeopardy if a roof fails. However, it is equally important that roof system selections be made in such a way as to assure that the best technologies and materials are used for each application. Often, what appears to be the safest choice is not in the best economic interest of the client or building owner.

In some cases, the problem is simple: the owner wants the system with the lowest initial cost due to budgetary constraints. However, many owners have found out the hard way that initial cost should not always be the most important determinant of a roof system. Factors such as longevity, maintainability, and resistance to high traffic or contaminants should all play a role in selection.

Design and Selection Parameters

Building Characteristics

Several questions should be answered before the appropriate roof system is chosen. As outlined in Chapter 2, many factors of the building itself, such as shape, height, roof slope, decking material, amount of roof insulation required, rainfall and drainage considerations, parapet details, and environment must be considered. As we have outlined in the past chapters, certain types of decks are more compatible with certain roofing systems and attachment methods. These factors should all be weighed, and a selection made based on objective criteria. The system selection should meet the most criteria possible. This chapter, as well as Chapters 2 and 4, present an overview of roofing and insulation selection. These guidelines should be helpful in selecting the optimal system for any application.

Early Stage Design Tradeoffs

Various factors can be changed early in the design process in order to improve the roof design. For example, the locations of mechanical, electrical, and plumbing penetrations, as well as skylights and hatches can be revised. Even some of the major items such as the number of corners on the building might be changed. These possibilities are covered in more detail later in this chapter.

Some roof systems are simply incompatible with certain deck or insulation types, or do not lend themselves to particular applications. Aesthetic considerations, attachment problems with a type of deck or insulation which cannot be changed, lack of local or area installers who have a background with that particular system — these are all factors that may preclude certain roof system types for any given locale. In some cases, design tradeoffs can be made with regard to slopes, deck material, insulation methods, or even building shape. For these reasons, it is advisable for specifiers to keep roof system selection in mind from the outset, in order to make structural, aesthetic, or functional decisions that will have a positive impact on the project and allow for the greatest versatility in roof material selection.

Locating Qualified Roofing Contractors

Owners may request that the specifier collaborate with a qualified roofing consultant. This consultant can aid in specifying the optimal roof system for the building, locate qualified installers, and recommend maintenance procedures. The National Roofing Contractors Association (NRCA) maintains listings of all its members in the area of roofing consultation. (See Appendix A for information on contacting the NRCA.)

In many instances, there is a perception of conflict of interest because a consultant may also be a roofing contractor. In these cases, if the consultant is experienced and professional, that person's firm may still be allowed to bid on the project. However, assurance must be given that the process of consultation and system selection is separated from that of bidding the project. Consultation should be remunerated so that the consultant does not feel that he has an "investment" in getting the job, and the owner/specifier feels no undue obligation to the bidder associated with the consultant.

One of the goals of this book is to give the architect, specifier, or facilities manager many of the tools available to roofing consultants so that the services of a consultant may not be required. In some cases, it may still be advisable to retain a consultant for projects involving complex designs or demanding conditions.

Sources of Contractor Lists

If a decision is made to use the services of a roofing consultant, this consultant will often be familiar with the most qualified roofers in the project area. The NRCA also maintains a list of its member contractors. (See Appendix A for information on NRCA's member hotline.)

The names of qualified local contractors can be obtained from the following sources: the NRCA; local builder's exchange or Contractor's Association; large facilities operators in the area, such as manufacturers, schools and colleges, and hospitals. This latter group normally has a list of "approved" roofers who have done satisfactory work for them in the past.

If the roof system has been selected prior to locating local installers, the manufacturer can be a good source of names of contractors. The manufacturer will have a list of its qualified installers in the project area, for the system chosen. In many cases, if the system is relatively new, such as a mechanically attached single-ply system, the manufacturer will not recommend any installers other than those who are trained and experienced in installing their product. Sometimes a competent local installer can be trained —partially in the factory and partially in the field — by a manufacturer's representative, with the representative having close control and scrutiny of the installation.

Qualified Contractor List

After consulting with all available sources of contractor names, a list of qualified roofers should be compiled. Those selected for the list should be judged on the basis of financial strength, years of experience in the area, and familiarity with the particular system specified. No fewer than three names should be on the list, which should be used in determining who will be allowed to bid the project. The list should be made public at the pre-bid conference in order to discourage unqualified contractors from bidding the job. Larger firms will often not bid a project on which any and all bidders are welcomed. Such prequalification measures assure that the contract administrator will not be put in the position of having to choose between a "lowball" price and a qualified installer.

Roof System Selection Chart

The Roof System Selection Chart (Figure 17.1) has been developed based on the author's own experience and various industry sources of information. It is not meant to be a comprehensive list of all system possibilities, nor to preempt the specifier's own professional judgment.

In the selection process mentioned earlier, there may be factors that indicate the use of a particular, possibly proprietary, system. This system, in the specifier's sole judgment, is the best suited to a particular project's design parameters. In most cases, however, the selection process is hardly that clear-cut. The designer or specifier should use all sources of information, such as industry journals, technical papers, and information releases from organizations such as NRCA and others. The best guide to whether or not a system is really as good as it is advertised, is simple: how many installations exist, and how long have they been in service? What is the service record for installations similar to the one proposed for this project?

Outlined here are the characteristics listed in the chart for each generic roofing system. The roofing systems listed are numerically rated for various characteristics. Local conditions, or the specifier's own experience may lead to a conclusion different from that drawn

Roof System Selection Chart

Note: This is not intended to be a comprehensive listing of available generic systems, but an overview of the largest and/or fastest growing segments of the roofing industry as of the time of publication.

ROOF SYSTEM TYPE	CHARACTERISTICS	Low initial cost	Many re-entrant corners & penetrations	High traffic roofs or areas	Resistant to oils, solvents	Mid-rise or high-rise use (over 5 stories)	Ease of maintenance	Compatible with light steel decks	Long record of serviceability	REMARKS
Built-up roof, Asphalt		3	5	5	3	1	5	1	5	Traditional materials.
Built-up roof, Coal Tar Pitch		2	4	5	3	1	5	1	5	Traditional materials.
Modified Bitumen (APP modifier)		2	5	4	4	3	4	2	3	Torch-applied material.
Modified Bitumen (SBS modifier)		2	5	4	4	3	5	2	3	Mop-applied material.
Single-ply, ballasted*		4	2	2	2	1	2	1	3	Simple & inexpensive.
Single-ply, adhered*		2	4	2	2	4	3	4	3	High growth in this segment.
Single-ply, mechanically attached*		3	3	2	2	4	3	4	3	Rapid development in this segment.
Non-vulcanized elastomers, mechanically attached		2	2	2	4	4	3	4	3	Relatively small market segment.
Preformed Metal		2	2	2	4	2	2	3	5	Durable material.
Spray-applied Polyurethane		2	5	1	2	3	1	2	1	Requires training and careful quality control.

*Characteristics listed are for EPDM single-ply sheet; other elastomers have similar characteristics.

A rating of: 1 = poor; not recommended
2 = fair
3 = average
4 = better than average
5 = excellent; recommended

Figure 17.1

in the chart. The chart is intended as a general guide, and is established from industry information on systems in use.

Initial Cost

The factor of initial cost is actually a complex one. Although a low installation cost is a priority for the owner, it should be just one of several considerations to the specifier. While ballasted or mechanically-attached EPDM, or asphaltic BUR may seem to be the cheapest systems up-front, they may not be suited to a particular project for other reasons.

Roof Penetrations and Building Design

In buildings that contain many angles, corners, penetrations, and pieces of rooftop equipment, certain types of systems are not recommended, due to the relative difficulty in constructing turns and flashings. The designer is well advised to weigh the benefits versus the costs of an extremely "cut-up" design; roofing is only one of the many systems that are adversely affected by increasing building complexity.

Mechanical and electrical penetrations can often be grouped together in order to reduce the number of room penetrations. Coordination of such basic items is a sign of a thoughtful architect. Mechanical and electrical engineers are often surprised to see this level of coordination in the profession. Indeed, everyone benefits from cooperation among design professionals. There is an even more direct advantage to the architect who gives thought to roofing as it affects building envelope design: a "clean" roof greatly increases the aesthetic appeal of a building, and makes it more serviceable over the long run.

When the building design makes a "cut-up" roof unavoidable, the BUR and modified bitumen systems are the strongest choice, depending on other factors, such as building height and the type of deck. Spray-applied polyurethane is suited to odd-shaped applications, but the difficulty of proper application and careful quality control requirements may preclude its use.

High Traffic Roofs

A roof used as a promenade deck requires a tough membrane, or protection, such as pavers or an inverted roof assembly. In any case, care should be taken to install walkways and traffic pads which are approved by the membrane manufacturer, in order to protect the membrane from traffic damage or tool droppage. Polyurethane spray-applied roofing and the single-ply systems are the least resistant to traffic damage and physical abuse, whereas the BUR and modified systems tend to have a "self-healing" tendency and are more resistant to surface damage due to their highly resilient surface. This is not to suggest, however, that one can afford to forego the requirement for traffic pads or walkways.

Contaminants: Chemical Damage

Environmental factors are critical in material selection. For example, the roof of an airport terminal building might require an aesthetically pleasing membrane with resistance to jet exhaust fumes, while a restaurant or commercial kitchen roof will require that the membrane not be adversely affected by airborne grease from exhaust hoods. Chemical laboratories and manufacturing or refining process buildings require a roof that will be resistant to chemicals found in their exhaust stack emissions.

Within the single-ply roof system category, as is evidenced in the Single-Ply Comparison Chart in Chapter 8, each SPR material is affected by different chemicals and contaminants. The Roofing Questionnaire, a sample of which is included in Appendix B of this book, may be helpful in conducting this analysis.

The Roofing Questionnaire

The Roofing Questionnaire (Appendix B), when used during the preliminary specifier/owner meetings, will provide the specifier with important information about the owner's operational environment that may affect the roof selection. Conducting this interview with the client early in the relationship is one more way in which the architect may show concern about the often overlooked operational and practical aspects of the building.

Tall Buildings

In most areas, mid-rise or high-rise buildings (over ten stories) preclude the use of both ballasted SPR systems (due to blow-off in high winds) and BUR systems (due to the impracticality of working with hot bitumens and other BUR materials on a high-rise site).

Ease of Maintenance

The following characteristics are worth noting regarding roof maintenance. First, coated spray-applied polyurethane roofing has had a poor maintenance history, largely due to improper application or coating. This material is used for a relatively small segment of the roofing market. It has characteristics which make it the ideal choice for difficult shapes and complex surfaces. For most installations, however, there are other selections that are far easier to maintain.

Second, ballasted SPR systems can be difficult to maintain due to the difficulty of locating leaks. This is because the ballast hides the evidence of failure. Once the leak is discovered, however, it is relatively easy to repair, unless the system is an inverted membrane.

Light Steel Decks

For use on light steel decks, which represent a large number of today's commercial building roof decks, the SPR systems (except for ballasted systems) are the best choice. Ballasted systems and built-up roofs are either too heavy to be accommodated by light roof structures, or are not compatible with the flexure and temperature deformation of light steel decks.

Long Service Record

Metal roofing systems and built-up roofs have the longest service record of any of the major systems used in commercial buildings today. Tile, wood shingles, slate, terneplate, and other traditional materials, while well known for over a hundred years, are not used in any significant quantity in today's commercial roofing market.

Although sprayed polyurethane foam (PUF) has been in service as a roofing material for many years, and as an insulating material for 40 years, early formulations exhibited serious shrinkage and deterioration problems. Sprayed polyurethane, as it is used today, represents a relatively small percentage of the total roof market.

Regulatory Considerations

Building codes and the requirements of Factory Mutual or Underwriters' Laboratory may preclude the use of certain systems for a given application. For example, some codes restrict the use of kettles on the roof, thereby making built-up systems impractical. Factory Mutual's stringent standards for fastener type, holding requirements, and spacing sometimes make certain systems prohibitively expensive in windy areas.

Fire codes now prevent the use of torches in some areas for roofing applications. As a result, modified bitumen roofing using APP modifiers is not allowed.

Summary

The architect or specifier's choice of roofing system may be one of the most important decisions on a project. Not only is the roof one of the biggest cost items in most structures, but it is also one of the most likely components to require repair. Roof selection should, therefore, be an objective, careful, and thorough process, based on all applicable information. Crucial to the selection process are early planning and coordination among the trades, and obtaining qualified roofing contractors.

CHAPTER EIGHTEEN

ESTIMATING

CHAPTER EIGHTEEN

ESTIMATING

Types of Estimates

Construction estimators use four basic types of estimates. These types may be referred to by different names and may not be recognized by all as definitive, but most estimators will agree that each type has its place in the construction estimating process. Figure 18.1 shows the four basic types of estimates, characterized by the preparation time required versus the accuracy of the method.

1. *Order of Magnitude Estimate:*
   The order of magnitude estimate could be loosely described as an educated guess. It is generally used at the earliest stage of project development when the building type and use are decided. An order of magnitude estimate can be completed in a matter of minutes. The accuracy range is plus or minus 20%.

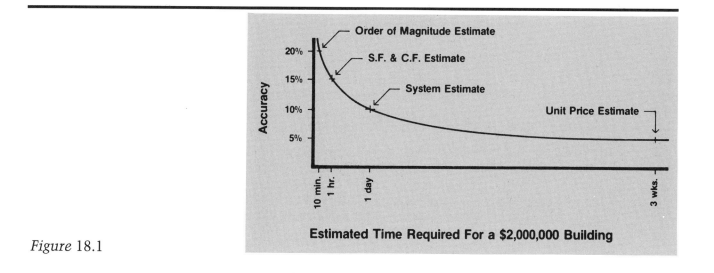

Figure 18.1

2. *Square Foot Estimate:*
   This type of estimate is most often used when the proposed size and function of a planned building are known, but plans have not yet been prepared. Very little information is required for this basic estimate, and accuracy is plus or minus approximately 15%.

3. *Systems (Assemblies) Estimate:*
   A systems estimate is best used as a budgetary tool in the planning stages of a project. Accuracy is expected at plus or minus 10%.

4. *Unit Price Estimate:*
   Working drawings and full specifications are required to complete a unit price estimate. It is the most accurate of the four types, but it is by far the most time consuming. Used primarily for bidding purposes, the accuracy of a unit price estimate is plus or minus 5%.

Tools for Estimate Preparation

The various types of estimates require different tools, ranging from historical cost data from the estimator's own files (which might also include quotes from suppliers and subcontractors), to published sources such as Means' annual cost data manuals, to automated methods, such as a roof estimate spreadsheet generated on a microcomputer or personal computer. For computerized estimate preparation, the database may consist of information derived from the estimator's own experience and suppliers, or it may come from a published source, such as Means. An example of a computer spreadsheet, using a hypothetical project and prices, is illustrated in Figure 18.2. The building in this project has two types of roofing: ballasted EPDM (Type "A") and clay mission tile (Type "B"). The first three columns of the spreadsheet show the estimated quantity, units, and item description. The next two show the cost per unit for materials and labor, which should come from the estimator's files on similar recent projects. The last two columns are the totals for these quantities. At the bottom of Figure 18.2, the spreadsheet completes the roofing estimate by computing the additional direct and indirect costs involved in this project.

For a roofing project, terms such as *man-hours per square* and *bid price per square* can be added to internal calculations in the estimate spreadsheet, so that the print-out will alert the estimator to an overly low or high number.

Computerized Estimating

There are numerous spreadsheet systems on the market today for use with personal computers. In addition to computer-aided design and drafting (CADD) and many other office functions, the computer is playing an ever-larger role in architectural and engineering firms. Contractors are beginning to realize increased efficiency through computer use and, consequently, a competitive "edge." There are many spreadsheet programs, some integrated in larger packages along with word processing and other business functions. The best way to select the most suitable program is to visit a software vendor or computer store, explain the firm's requirements, and allow them to demonstrate how the firm's particular estimate sheets may best be integrated into a spreadsheet format.

ROOFING ESTIMATE

PROJECT NAME: Plymouth Manufacturing Co.
DATE OF ESTIMATE: Nov. 19, 1988
LOCATION: Plymouth, MA
USE: Light Manufacturing & Assembly, Electronics
TOT. ROOF AREA: 52,000 s.f.
ROOF TYPE "A":   37,500 s.f. (Ballasted EPDM)
ROOF TYPE "B":   14,500 s.f. (Clay Mission Tile)
ROOF TYPE "C":   n/a
ROOF TYPE "D":   n/a

ROOF TYPE A

QTY.	UNITS	DESCRIPTION	COST PER UNIT MAT'L	LABOR	TOTAL MAT'L.	TOTAL LABOR
37500	SF	45 MIL EPDM	$0.49	$0.11	$18375.00	$4125.00
12000	LF	SEAM PREP	0.01	0.06	120.00	720.00
12000	LF	LAP SEAL	0.03	0.09	360.00	1080.00
3450	LF	EPDM 12" FLASH	0.51	0.60	1759.50	2070.00
185	TONS	BALLAST (STONE)	15.00	21.00	2775.00	3885.00
160	LF	EQPT. CURB	4.50	2.25	720.00	360.00
775	LF	EXPAN. JT.	3.75	1.25	2906.25	968.75
310	LF	AREA DIVIDER	2.25	1.05	697.50	325.50
12	EA.	PENETRATIONS		6.75		81.00
28	EA.	ROOF DRAINS		1.75		49.00
36	LF	SCUPPERS		5.00		180.00
		TOTAL FOR TYPE "A"			$27713.25	$13844.25

ROOF TYPE B

QTY.	UNITS	DESCRIPTION	COST PER UNIT MAT'L	LABOR	TOTAL MAT'L	TOTAL LABOR
14500	SF	CLAY TILE	$5.10	$1.35	$73950.00	$19575.00
310	LF	FLASHING,COPPER	1.85	1.25	573.50	387.50
310	LF	REGLET, COPPER	1.88	0.75	582.80	232.50
310	LF	COUNTERFLASHING	1.50	1.28	465.00	396.80
310	LF	GUTTER, COPPER	3.15	1.60	976.50	496.00
		TOTAL FOR TYPE "B"			$76574.80	$21087.80

LABOR TOTAL	$34,932.05
MAT'L TOTAL	$104,261.05
TAX	$9,743.51
TOTAL DIRECT COSTS	$148,936.61
WARRANTY RESERVE	$7,550.00
PERMITS, FEES	$875.00
BONDS	$10,000.00
OVERHEAD (20%)	$29,780.00
TOTAL COST	$197,141.61
PROFIT (10%)	$19,715.00
ESTIMATE TOTAL	$216,856.61

Figure 18.2

The potential benefits of computerized spreadsheets, in terms of organization and time savings, can revolutionize a roofing operation, especially if the company is already short on qualified administrative personnel. After the initial learning period, which can be extremely short in the case of some of the more user-friendly systems, many designers and contractors find an automated system indispensable.

Traditional Estimating

Possibly the most important estimating tool, whatever method is chosen, is the estimator's own experience and knowledge of the local market in terms of expected weather delays, costs of materials and labor, degree of competition, and expected difficulty of the job. The traditional approach to estimating is to use a variety of local sources of information to arrive at a reliable estimate price.

Using Published Cost Data

National average price information on roofing items and complete installations can be obtained from published sources, such as R.S. Means' annually updated cost books. The City Cost Index included in these publications provides a factor for converting Means national average prices to a specific locality. Using these cost guides can make systems or unit price estimates far simpler, and help to remind the estimator of often-forgotten items of work or materials. This approach to systems and unit price estimating is covered in more detail in the following sections.

Order of Magnitude Estimates: The order of magnitude estimate can be completed with only a minimum of information. The proposed use and size of the planned structure should be known and may be the only requirement. This information provides a preliminary indication of the type of roof system to be used. If possible, it is helpful to know the approximate roof slope and configuration of the structure. The "units" can be very general, and need not be well defined. For example, the following statement (or estimate) might be made after a few minutes of thought used to draw upon experience and to make comparisons with similar projects from the past. "The roofing for a large administrative facility for a service company, in a suburban industrial park, will cost about $250,000." While this rough figure might be appropriate for a project in one region of the country, an adjustment may be required for a change of location and for cost changes over time.

Figure 18.3, a page from the 1989 edition of R.S. Means' *Building Construction Cost Data*, illustrates examples of a different approach to the order of magnitude estimate. This format is based on *unit of use*. Note that at the bottom of the categories "Hospitals" and "Housing," costs are given "per bed" or "person," and "per rental unit" and "per apartment." This data does not require that details of the proposed project be known in order to determine rough costs; the only required information is the intended use of the building and its approximate size. What is lacking in accuracy is offset by the minimal time required to complete the order of magnitude estimate, a matter of minutes.

Square Foot Estimates: The square foot estimate is most appropriate prior to the preparation of project drawings, when the project budget is being established. Note that in Figure 18.3, costs

171 000	S.F. & C.F. Costs		UNIT COSTS			% OF TOTAL			
		UNIT	¼	MEDIAN	¾	¼	MEDIAN	¾	
410 9000	Per car, total cost	Car	5,375	7,325	10,300				410
9500	Total: Mechanical & Electrical	"	385	580	710				
430 0010	GYMNASIUMS	S.F.	44.10	59.35	76				430
0020	Total project costs	C.F.	2.20	3.06	3.90				
1800	Equipment	S.F.	1.14	1.90	3.33	2%	3.20%	6.70%	
2720	Plumbing		2.72	3.70	4.68	4.80%	7.20%	8.50%	
2770	Heating, ventilating, air conditioning		2.89	5	8.20	7.40%	9.70%	14%	
2900	Electrical		3.71	4.53	6.75	6.50%	8.90%	10.70%	
3100	Total: Mechanical & Electrical	↓	8	12.45	16.35	16.70%	21.80%	27%	
3500	See also division 114-801 & 114-805								
460 0010	HOSPITALS	S.F.	97.40	118	159				460
0020	Total project costs	C.F.	7.15	8.60	11.50				
1120	Roofing	S.F.	.71	1.77	2.88	.50%	1.20%	2.90%	
1320	Finish hardware		.96	1.04	1.26	.60%	1%	1.20%	
1540	Floor covering		.67	1.22	3.07	.50%	1.10%	1.60%	
1800	Equipment		2.40	4.44	6.64	1.60%	3.80%	5.60%	
2720	Plumbing		8.60	10.95	15	7.50%	9.10%	10.80%	
2770	Heating, ventilating, air conditioning		9.30	15.50	21.75	8.40%	13%	16.70%	
2900	Electrical		10.05	13.45	20.45	10%	12.30%	15.10%	
3100	Total: Mechanical & Electrical	↓	29.45	40.15	59.35	28.20%	36.60%	40.30%	
9000	Per bed or person, total cost	Bed	26,400	51,400	67,700				
9900	See also division 117-001								
480 0010	HOUSING For the Elderly	S.F.	47.35	59.40	74.50				480
0020	Total project costs	C.F.	3.34	4.63	6.05				
0100	Sitework	S.F.	3.38	5.10	7.35	6%	8.20%	12.10%	
0500	Masonry		1.42	4.98	7.60	2.10%	6.50%	10.40%	
0730	Miscellaneous metals		1.39	1.85	4.88	1.40%	2.10%	3.10%	
1120	Roofing		.92	1.59	2.81	1.20%	2.10%	3%	
1140	Dampproofing		.27	.37	.85	.20%	.40%	.70%	
1340	Windows		.64	1.09	2.35	1.10%	1.50%	2.40%	
1350	Glass & glazing		.17	.49	.91	.20%	.40%	.90%	
1530	Drywall		2.81	3.69	5.91	3.70%	4.10%	4.60%	
1540	Floor covering		.76	1.13	1.63	.90%	1.40%	1.90%	
1570	Tile & marble		.36	.52	.78	.50%	.60%	.80%	
1580	Painting		1.52	2.14	3.32	2%	2.60%	3.10%	
1800	Equipment		1.09	1.51	2.39	1.80%	3.20%	4.40%	
2510	Conveying systems		1.11	1.52	2.02	1.70%	2.30%	2.80%	
2720	Plumbing		3.58	4.96	7.24	8.30%	9.70%	10.90%	
2730	Heating, ventilating, air conditioning		1.60	2.48	3.50	3.20%	5.60%	7.10%	
2900	Electrical		3.48	4.87	6.85	7.50%	9%	10.60%	
2910	Electrical incl. electric heat		3.98	7.50	8.90	9.60%	11%	13.30%	
3100	Total: Mechanical & Electrical	↓	8.65	12.20	15.90	18.40%	21.90%	24.70%	
9000	Per rental unit, total cost	Unit	41,800	49,500	54,400				
9500	Total: Mechanical & Electrical	"	8,300	10,400	12,300				
500 0010	HOUSING Public (low-rise)	S.F.	36.45	49.90	67.90				500
0020	Total project costs	C.F.	3.01	3.92	4.95				
0100	Sitework	S.F.	4.77	6.70	10.35	8.30%	11.70%	16.10%	
1800	Equipment		1.04	1.69	2.76	2.20%	3.10%	4.60%	
2720	Plumbing		2.60	3.59	4.61	7.10%	9%	11.50%	
2730	Heating, ventilating, air conditioning		1.39	2.60	2.95	4.40%	6%	6.40%	
2900	Electrical		2.29	3.29	4.62	5%	6.50%	8.10%	
3100	Total: Mechanical & Electrical	↓	6.80	9.70	13.55	15.60%	19.20%	22.20%	
9000	Per apartment, total cost	Apt.	39,700	45,000	56,200				
9500	Total: Mechanical & Electrical	"	6,675	9,150	11,500				
510 0010	ICE SKATING RINKS	S.F.	33.90	47.35	77.70				510
0020	Total project costs	C.F.	1.92	2.41	2.84				
2720	Plumbing	S.F.	1.02	1.50	2.29	3.10%	3.20%	4.60%	
2900	Electrical	"	2.67	3.51	4.87	5.70%	7%	10.10%	

362   For expanded coverage of these items see *Means Square Foot Cost Data 1989*

Figure 18.3

for each type of project are presented first as "total project costs" per square foot and by cubic foot. These costs are broken down into different components, and then into the relationship of each component to the project as a whole, in terms of cost per square foot. This breakdown enables the designer, planner, or estimator to adjust certain components according to the unique requirements of the proposed project.

The best source of square foot costs is the estimator's own cost records for similar projects, adjusted to the requirements of the project in question. While helpful for preparing preliminary budgets, square foot estimates can also be useful as checks against other, more detailed estimates. Square foot estimates require slightly more time than order of magnitude estimates, but a greater accuracy (plus or minus 15%) is achieved due to more specific definition of the project.

Systems (or Assemblies) Estimates: Rising design and construction costs have made budgeting and cost efficiency increasingly important in the early stages of building projects. Never before has the estimating process had such a crucial role in the initial project planning. Unit price estimating, because of the time and detailed information required, is not the most appropriate budgetary or planning tool. A faster and more cost effective method for the planning phase is the systems, or assemblies estimate.

The systems method is a logical, sequential approach that reflects the order in which a building is constructed. *Means Assemblies Cost Data* lists the labor and material costs of construction items according to this *Uniformat* organization. Twelve "Uniformat" divisions organize building construction into major components. These Uniformat divisions are listed below:

- •Division 1 Foundations
- •Division 2 Substructures
- •Division 3 Superstructure
- •Division 4 Exterior Closure
- •Division 5 Roofing
- •Division 6 Interior Construction
- •Division 7 Conveying
- •Division 8 Mechanical
- •Division 9 Electrical
- •Division 10 General Conditions
- •Division 11 Special
- •Division 12 Site Work

Each division is further broken down into *systems*. Each of these systems incorporates all of the components required to install, for example, a built-up roof system. With a complete system cost there is no need to refer to several areas of the unit price publications in order to price each item, such as the plies of felt, the asphalt or coal tar, the gravel, and the base sheet. Figure 18.4 is an example of a typical Means system, in this case a *Built-up Roof Cover* from the 1989 edition of *Means Assemblies Cost Data*.

A great advantage of the systems estimate is that the estimator's specifier is able to substitute one roofing system for another during design development and quickly determine the cost differential.

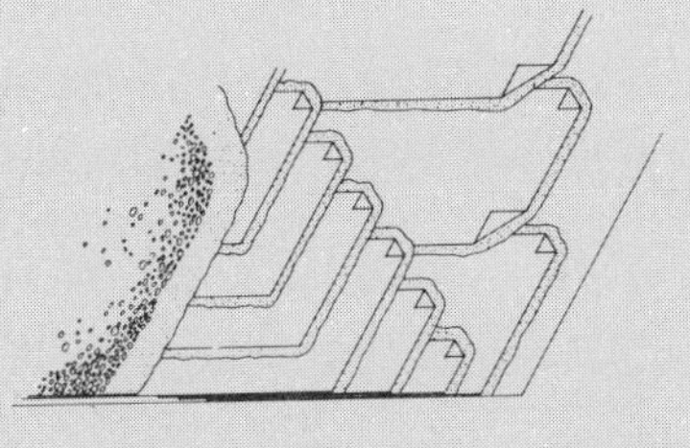

Multiple ply roofing is the most popular covering for minimum pitch roofs. Lines 1200 through 6300 list the costs of the various types, plies and weights per S.F.

Systems Components	QUANTITY	UNIT	COST PER S.F.		
			MAT.	INST.	TOTAL
SYSTEM 05.1-102-2500					
ASPHALT FLOOD COAT, W/GRAVEL, 4 PLY ORGANIC FELT					
Organic #30 base felt	1.000	S.F.	.06	.04	.10
Organic #15 felt, 3 plies	3.000	S.F.	.09	.13	.22
Asphalt mopping of felts	4.000	S.F.	.12	.37	.49
Asphalt flood coat	1.000	S.F.	.08	.32	.40
Gravel aggregate, washed river stone	4.000	Lb.	.03	.07	.10
TOTAL			36.32	85.98	122.30

5.1-103	Built-up	COST PER S.F.		
		MAT.	INST.	TOTAL
1200	Asphalt flood coat w/gravel; not incl. insul, flash., nailers			
1300				
1400	Asphalt base sheets & 3 plies #15 asphalt felt, mopped	.39	.81	1.20
1500	On nailable deck	.42	.88	1.30
1600	4 plies #15 asphalt felt, mopped	.46	.89	1.35
1700	On nailable deck	.42	.93	1.35
1800	Coated glass base sheet, 2 plies glass (type IV), mopped	.38	.82	1.20
1900	For 3 plies	.46	.89	1.35
2000	On nailable deck	.42	.93	1.35
2100	3 plies glass fiber felt (type IV), mopped	.46	.84	1.30
2200	On nailable deck	.42	.88	1.30
2300	4 plies glass fiber felt (type IV), mopped	.54	.91	1.45
2400	On nailable deck	.50	.95	1.45
2500	Organic base sheet & 3 plies #15 organic felt, mopped	.39	.91	1.30
2600	On nailable deck	.36	.94	1.30
2700	4 plies #15 organic felt, mopped	.43	.82	1.25
2750				
2800	Asphalt flood coat, smooth surface			
2850				
2900	Asphalt base sheet & 3 plies #15 asphalt felt, mopped	.37	.73	1.10
3000	On nailable deck	.34	.76	1.10
3100	Coated glass fiber base sheet & 2 plies glass fiber felt, mopped	.35	.70	1.05
3200	On nailable deck	.33	.77	1.10
3300	For 3 plies, mopped	.42	.78	1.20
3400	On nailable deck	.39	.81	1.20
3500	3 plies glass fiber felt (type IV), mopped	.43	.72	1.15
3600	On nailable decks	.40	.75	1.15

For expanded coverage of these items see *Means Building Construction Cost Data 1989*

192

Figure 18.4

The owner can then anticipate more accurate budgetary requirements before final sizes, details, and systems are established. The time and effort required for detailed design and specification have not yet been invested, and the design may still be changed without great expense. The plus or minus 10% accuracy range is normally close enough to satisfy the owner's requirement at this point.

The systems method does not require detailed information, but estimators who use it must have a solid background knowledge of roofing materials and methods, Building Code requirements, design options, and budgetary restrictions.

The systems estimate, unlike the unit price method, should not be used as a substitute for the unit price estimate. While it can be an invaluable tool in the planning stages of a project, it should be supported by unit price estimating when greater accuracy is required, and the design is in the final stages.

Unit Price Estimates: The unit price estimate is the most accurate and detailed of the four estimate types and therefore requires the most time to complete. Detailed working drawings and specifications must be available to the unit price estimator. All decisions regarding the roofing system's materials and methods must have been made previously in order to complete this type of estimate. With fewer variables than the estimate types discussed previously, accuracy is greatly enhanced. Labor and material quantities are determined from the working drawings and specifications, as are equipment requirements. Current and accurate costs for these items (in the form of unit prices) are also necessary. These costs may come from different sources, but whenever possible, should be based on experience or cost figures from similar projects. If the estimator does not have his own historical cost data available, prices may be determined instead from an up-to-date industry source book such as Means' *Building Construction Cost Data.*

Because of the detail involved and the need for accuracy, the preparation of unit price estimates requires a great deal of time and expense. For this reason, unit price estimating is best suited for the preparation of project bids. It can also be effective for determining certain detailed costs in conceptual budgets or during design development.

Most construction specification manuals and cost reference books, such as Means' *Building Construction Cost Data* organize all unit price information according to the 16 Construction Specifications Institute's MASTERFORMAT divisions, listed below.

- •Division 1 General Requirements
- •Division 2 Site Work
- •Division 3 Concrete
- •Division 4 Masonry
- •Division 5 Metals
- •Division 6 Wood and Plastics
- •Division 7 Moisture — Thermal Control
- •Division 8 Doors, Windows & Glass
- •Division 9 Finishes
- •Division 10 Specialties

- Division 11 Equipment
- Division 12 Furnishings
- Division 13 Special Construction
- Division 14 Conveying Systems
- Division 15 Mechanical
- Division 16 Electrical

This method of organizing construction components provides a standard of uniformity widely used by construction industry professionals: architects, engineers, material suppliers, and contractors. A sample page from the 1989 edition of Means' *Building Construction Cost Data* is shown in Figure 18.5. Each page contains a wealth of information useful in unit price estimating. The type of work to be performed is described in detail, including typical roof crew make-up, daily output, and separate costs for material and installation. Total costs are extended to include the installing contractor's overhead and profit. Figure 18.6 is an example of a handwritten unit price roofing estimate. The prices used in these estimates are from Means 1989 cost data. The estimate forms are from *Means Forms for Building Construction Professionals.*

Summary

Estimating for roof installation or replacement is a key function in the successful project. This task is made easier and more accurate if the estimator maintains a file of historical cost data, based on past projects and current prices from local suppliers and subcontractors. The estimator's own records, and each estimate, should also include appropriate costs for warranties, maintenance, and inspections.

076 200	Sheet Mtl Flash & Trim	CREW	DAILY OUTPUT	MAN-HOURS	UNIT	BARE COSTS MAT.	LABOR	EQUIP.	TOTAL	TOTAL INCL O&P	
201	5800	3" x 4"	1 Shee	145	.055	L.F.	1.30	1.33		2.63	3.43	201
	6000	Rectangular, plain, 28 gauge, galvanized, 2" x 3"		190	.042		.45	1.01		1.46	2.02	
	6100	3" x 4"		145	.055		1.30	1.33		2.63	3.43	
	6300	Epoxy painted, 24 gauge, corrugated, 2" x 3"		190	.042		.80	1.01		1.81	2.40	
	6400	3" x 4"		145	.055		1.37	1.33		2.70	3.50	
	6600	Wire strainers, rectangular, 2" x 3"		145	.055	Ea.	1.40	1.33		2.73	3.54	
	6700	3" x 4"		145	.055		2.25	1.33		3.58	4.47	
	6900	Round strainers, 2" or 3" diameter		145	.055		.95	1.33		2.28	3.04	
	7000	4" diameter		145	.055		1.20	1.33		2.53	3.32	
	7200	5" diameter		145	.055		1.80	1.33		3.13	3.98	
	7300	6" diameter		115	.070		2.20	1.67		3.87	4.94	
	7500	Steel pipe, black, extra heavy, 4" diameter		20	.400	L.F.	11.25	9.60		20.85	27	
	7600	6" diameter		18	.444		24	10.70		34.70	42	
	7800	Stainless steel tubing, schedule 5, 2" x 3" or 3" diameter		190	.042		8.60	1.01		9.61	11	
	7900	3" x 4" or 4" diameter		145	.055		10.75	1.33		12.08	13.80	
	8100	4" x 5" or 5" diameter		135	.059		17.70	1.43		19.13	22	
	8200	Vinyl, rectangular, 2" x 3"		210	.038		.87	.92		1.79	2.34	
	8300	Round, 2-½"		220	.036		.87	.87		1.74	2.27	
202	0010	DRIP EDGE Aluminum, .016" thick, 5" girth, mill finish	1 Carp	400	.020		.15	.43		.58	.81	202
	0100	White finish		400	.020		.17	.43		.60	.83	
	0200	8" girth		400	.020		.30	.43		.73	.97	
	0300	28" girth		100	.080		1.20	1.71		2.91	3.90	
	0400	Galvanized, 5" girth		400	.020		.22	.43		.65	.89	
	0500	8" girth		400	.020		.26	.43		.69	.93	
203	0010	ELBOWS Aluminum, 2" x 3", embossed	1 Shee	100	.080	Ea.	.53	1.92		2.45	3.48	203
	0100	Enameled		100	.080		.53	1.92		2.45	3.48	
	0200	3" x 4", .025" thick, embossed		100	.080		.85	1.92		2.77	3.83	
	0300	Enameled		100	.080		.84	1.92		2.76	3.82	
	0400	Round corrugated, 3", embossed, .020" thick		100	.080		.77	1.92		2.69	3.74	
	0500	4", .025" thick		100	.080		.91	1.92		2.83	3.90	
	0600	Copper, 16 oz. round, 2" diameter		100	.080		4.25	1.92		6.17	7.55	
	0700	3" diameter		100	.080		5.30	1.92		7.22	8.75	
	0800	4" diameter		100	.080		9.25	1.92		11.17	13.05	
	0900	5" diameter		100	.080		11.80	1.92		13.72	15.90	
	1000	2" x 3" corrugated		100	.080		9	1.92		10.92	12.80	
	1100	3" x 4" corrugated		100	.080		12.70	1.92		14.62	16.85	
	1300	Vinyl, 2-½" diameter, 45° or 75°		100	.080		2.30	1.92		4.22	5.45	
	1400	Tee Y junction		75	.107		12.30	2.57		14.87	17.40	
204	0010	FLASHING Aluminum, mill finish, .013" thick		145	.055	S.F.	.28	1.33		1.61	2.30	204
	0030	.016" thick		145	.055		.33	1.33		1.66	2.36	
	0060	.019" thick		145	.055		.66	1.33		1.99	2.72	
	0100	.032" thick		145	.055		.80	1.33		2.13	2.88	
	0200	.040" thick		145	.055		1.35	1.33		2.68	3.48	
	0300	.050" thick		145	.055		1.63	1.33		2.96	3.79	
	0400	Painted finish, add					.15			.15	.17	
	0500	Fabric-backed 2 sides, .004" thick	1 Shee	330	.024		.38	.58		.96	1.30	
	0700	.016" thick		330	.024		.95	.58		1.53	1.92	
	0750	Mastic-backed, self adhesive		460	.017		1.97	.42		2.39	2.80	
	0800	Mastic-coated 2 sides, .004" thick		330	.024		.44	.58		1.02	1.36	
	1000	.005" thick		330	.024		.55	.58		1.13	1.48	
	1100	.016" thick		330	.024		.97	.58		1.55	1.94	
	1300	Asphalt flashing cement, 5 gallon				Gal.	6.50			6.50	7.15	
	1600	Copper, 16 oz., sheets, under 6000 lbs.	1 Shee	115	.070	S.F.	2.60	1.67		4.27	5.40	
	1700	Over 6000 lbs.		155	.052		2.60	1.24		3.84	4.73	
	1900	20 oz. sheets, under 6000 lbs.		110	.073		3.25	1.75		5	6.20	
	2000	Over 6000 lbs.		145	.055		3.25	1.33		4.58	5.55	
	2200	24 oz. sheets, under 6000 lbs.		105	.076		3.90	1.83		5.73	7.05	
	2300	Over 6000 lbs.		135	.059		3.90	1.43		5.33	6.45	

Figure 18.5

CONSOLIDATED ESTIMATE

Division 7

PROJECT **Office Building**
LOCATION
CLASSIFICATION
SHEET NO. **1 of 2**
ESTIMATE NO. **89-1**
DATE **1989**

TAKE OFF BY **ETW** QUANTITIES BY **ETW** ARCHITECT PRICES BY **MG** EXTENSIONS BY **JW** CHECKED BY **TW**

DESCRIPTION	NO.	DIMENSIONS			QUANTITIES	UNIT	MATERIAL UNIT COST	MATERIAL TOTAL	LABOR UNIT COST	LABOR TOTAL	EQUIPMENT UNIT COST	EQUIPMENT TOTAL	TOTAL
Division 7: Moisture & Thermal Control													
Water Proofing													
Asphalt Coating	7-1	602	0600		3000	SF	0.14	0420	0.32	960			1380
Protective Board 1/4"	7-1	602	4000		3000	SF	0.35	1050	0.35	1050			2100
Vapor Barrier - 6 mil poly	7-1	922	0900		208	SQ.	1.70	354	4.63	963			1317
Insulation													
3 1/2" Fiberglass Ext.	7-2	118	0420		17,600	SF	0.27	4752	0.11	1936			6688
3 1/2" Fiberglass Int.	7-2	118	0820		7620	SF	0.20	1524	0.13	991			2515
Roof Deck 2 5/8" Urethane/Fiberglass	7-2	203	1300		18900	SF	0.96	18144	0.20	3780			21924
Roofing													
4-ply Built-up	7-5	102	0500		189	SQ	41.45	7835.00	52.00	9828	5.00	945	18608
4 x 4 Cant	7-5	103	0010		800	LF	0.55	440	0.49	392			832
Sub Totals								34519		19900		945	55364

Figure 18.6

CONSOLIDATED ESTIMATE

Division 7

SHEET NO. 2 of 2

ESTIMATE NO. 89-1

DATE 1989

CHECKED BY TW

PROJECT Office Building

LOCATION

CLASSIFICATION

ARCHITECT

TAKE OFF BY ETW QUANTITIES BY ETW PRICES BY MG EXTENSIONS BY JW

DESCRIPTION	NO	DIMENSIONS	QUANTITIES	UNIT	MATERIAL UNIT COST	MATERIAL TOTAL	LABOR UNIT COST	LABOR TOTAL	EQUIPMENT UNIT COST	EQUIPMENT TOTAL	TOTAL UNIT COST	TOTAL
Division 7 (Cont'd)												
Sheet Metal												
Gravel Stop		7.7 105 0350	800	LF	5.60	4480	1.43	1144				5624
Alum. Flashing		7.6 204 0100	200	SF	0.80	160	1.33	266				426
Accessories												
Roof Hatches		7.7 206 1200	2	EA	735	1470	99	198		7		1675
Smoke Vent		7.7 208 0200	1	EA	980.00	980	50.00	50				1030
Sub Total						7090		1658		7		8755
Sheet 1						34519		19900		945		55364
Sheet 2						7090		1658		7		8755
Division 7 Totals						41609		21558		952		64119

Figure 18.6 (cont.)

190

APPENDICES

Table of Contents

Roofing Organizations

American National Standards
Institute, Inc.
    1430 Broadway
    New York, NY 10018
    (212) 354-3300

American Society for Testing
and Materials
    1916 Race Street
    Philadelphia, PA 19103
    (215) 299-5400

American Society of Heating,
    Refrigeration, and Air-
    conditioning Engineers
    1791 Tullie Circle, N.E.
    Atlanta, GA 30329
    (404) 636-8400

Asphalt Institute
    Asphalt Institute Bldg.
    College Park, MD
    20740-1802
    (301) 277-4258

Asphalt Roofing Manufacturers
Association
    6288 Montrose Road
    Rockville, MD 20852
    (301) 231-9050

Factory Mutual Engineering
Corp.
    1151 Boston Providence
    Turnpike (Route 1)
    Norwood, MA 02062
    (617) 762-4300

Gypsum Association
    1603 Orrington Ave.
    Evanston, IL 60201
    (312) 491-1744

National Institute of Standard
Technology
    Center for Building
    Technology
    Gaithersburg, MD 20899
    (301) 975-2000

National Insulation
Contractors Association
    99 Canal Center Plaza
    Suite 222
    Alexandria, VA 22314
    (703) 683-6422

National Roofing Contractors
Association
    One O'Hare Center
    6250 River Road
    Rosemont, IL 60018
    (312) 318-6722
    Hotline: 1-800-USA-ROOF

National Roofing Foundation
    (See National Roofing
    Contractors Association)

National Roofing Legal
Resource Center
    (See National Roofing
    Contractors Association)

National Tile Roofing
Manufacturing Institute
    3127 Los Feliz Blvd.
    Los Angeles, CA 90039
    (213) 660-4411

Perlite Institute
    600 South Federal
    Chicago, IL 60605
    (312) 922-2062

Roofing Industry Educational
Institute
    7006 S. Alton Way, #B
    Englewood, CO 80112-2003
    (303) 770-0613

Society of Plastics Industry
    2400 East Devone
    Suite 301
    Des Plaines, IL 60018
    (312) 297-6150

Underwriters Laboratories, Inc
(UL)
    333 Pfingsten Road
    Northbrook, IL 60062
    (312) 272-8800

Roofing Questionnaire

The following minimum information should be obtained from the building owner's representative(s) in charge of facilities management, processes, and/or building use.

Answers to many of these questions regarding meteorological conditions are found in the records of the local or area National Weather Service Bureau office.

1. What is the average January temperature range for this location? August? What is the *interior* design temperature?

2. What is the average January humidity range, this location? August? What is the *interior* design humidity range?

3. What is the maximum hourly rate and absolute amount of the following forms of precipitation per 24 hour period? (rain, snow, sleet)

4. What is the largest size of hail normally occurring for this location in any average year?

5. What pollutants are common in this area, and in what amounts?

6. Are there an unusual number of birds in the area that may congregate on this roof? Insects? Rodents?

7. What are the average wind velocities for each month of the year? Maximum which can be expected in any year?

8. How many days of recorded sunshine (substantially clear) are recorded in the average year for this location?

9. What will the predominant activity be within the building? Are any special processes taking place?

10. Will any manufacturing, refining, combustion, or other similar processes take place in or on the premises? If so, what will be the primary composition of fumes or precipitants released?

11. Is there any activity on, or access required to, the roof? If so, what will be its nature?

12. What equipment will be located on the roof?

13. What is the expected height of the roof above surrounding grade?

14. What is the expected predominant curtain wall or parapet material?

15. Will there be a requirement for many penetrations, such as for mechanical or electrical vents, pipes, chases, conduits, ducts, and the like? If so, will they be grouped in one confined area, or widespread?

16. Are there any other special considerations or requirements that should be noted regarding the roof of this structure?

NRCA Deck Dryness Test

The following procedure, recommended by the NRCA, provides an acceptable means of testing the dryness of the roof deck:

1. Use approximately 1 pint of bitumen that is specified for use in the roof membrane, heated to a temperature that will ensure an application temperature of 400°F.

2. Pour the bitumen on the surface of the deck. If the bitumen foams, the deck is NOT dry enough to roof.

3. After the bitumen has cooled, an attempt should be made to strip the bitumen from the deck surface. If the bitumen strips clean from the deck, the deck is NOT dry enough to roof.

NRCA Venting Recommendations

A. General Information

In order to help obtain maximum service from roofing materials, some types of roof decks, insulation and fill materials need to be vented. The following venting recommendations are based on empirical evidence and actual roofing contractor field experience, and are considered by NRCA to be necessary practices. Venting should **not** be used to attempt to dry a wet system (see page 149).

B. Venting Locations

The following is a list of general guidelines for the locations at which roof vents should be installed.

1. All closed plenum, unvented roof systems should be vented topside by means of moisture relief vents.

2. Where the minimum dimension of a building is 50 feet or less, ventilation can generally be provided by the use of "edge" (perimeter) venting, employing an open type metal gravel guard, metal counterflashing or fascia. This procedure will, in most instances, permit cross-venting.

3. Where the minimum dimension of a building is greater than 50 feet, both "edge" venting and "field" venting (venting of the middle area of the roof) should be employed.

4. Vent stacks that are installed in the field portion of the roof should be placed in such a manner that one vent will vent approximately 10 roof squares (1,000 square feet) or less of roof area. On small roof areas, no fewer than two vents should be installed.

5. The size and location of the vents should satisfy the venting requirements of the job specifications. When venting provisions are not included in a roof system and it is believed that they may be necessary, the roofing contractor should notify the responsible party of the need for venting provisions.

   Although the venting procedures listed above are not always included in architectural specifications, they should be carefully considered and included for:

   a. Any cementitious, lightweight, insulating concrete fill poured over galvanized metal decks (both slotted and non-vented types), formboards or prestressed concrete decks.

   b. Any type of insulation installed over a vapor retarder.

   c. Any type of insulation installed over a building space that has high moisture content, such as swimming pools, laundries, paper mills, bottling plants, etc.

   d. Any type of insulation that is "spot-mopped" or "strip-mopped" to the substrate below.

   e. Any type of insulation that is applied over an existing roof membrane.

C. Moisture Relief Vents in the Field of the Roof

The following is a list of general guidelines for the type of moisture relief vents to install in the field portion of the roof.

1. Moisture relief vents for the field portion of the roof can be either "one-way"-type vents or open "two-way"-type vents.

2. If the average relative humidity is high, "one-way"-type vents should be employed. Some manufactured "one-way"-type vents employ a small flexible diaphragm, which will relieve outward pressure but will not allow moisture vapor to pass back into the vent and the roof insulation.

NRCA Venting (continued)

D. Installation Procedures for Moisture Relief Vents

The following is a list of general guidelines for the installation of moisture relief vents.

1. When vents are installed, the area beneath the vent should be cored out down to the top of the structural deck or the top of the vapor retarder (if a vapor retarder is incorporated into the roof system). This void should be filled with loose granular or fibrous insulating material.

2. The vent flange should be set in compatible mastic on top of the roof membrane and properly flashed (sealed) to the roof.

E. Spacing of Moisture Relief Vents

The diagram below shows the location of moisture relief vents spaced approximately 33 feet on center. This typical layout for the location of roof vents allows for one vent for every 10 roof squares (1,000 square feet) or less of roof area.

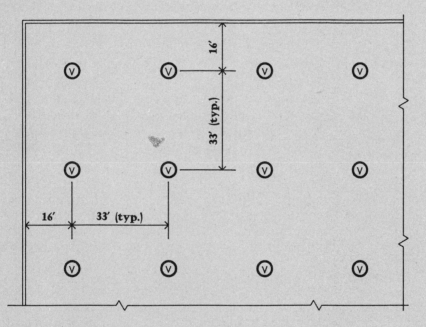

Typical Roof Layout of Moisture Relief Vents

General Guide to Fasteners

	Light-Weight Concrete Deck	Poured Gypsum Deck	Steel Deck	Wood Deck		Structural Plank Deck	
				Plywood Deck	Wood Plank Deck	Cement-Wood Fiber Deck	Precast Gypsum Concrete Deck
STANDARD ROOFING NAIL (3/8″-7/16″ diameter head) 11 or 12 gauge with barbed shank.					Use tin caps for nail heads smaller than 1″ dia.		
WEDGE NAIL Shank bends when driven (as shown) to provide back-out resistance.						Use tin caps for nail heads smaller than 1″ dia.	
THREADED ROOFING NAIL (3/8″ diameter head) Annular ring or spiral threaded, 11 gauge.				Use tin caps for nail heads smaller than 1″ dia.	Use tin caps for nail heads smaller than 1″ dia.		Use tin caps for nail heads smaller than 1″ dia.
SELF-LOCKING FASTENER (1″ diameter cap) Shank spreads out when driven to provide back-out resistance.							
TWO-PIECE TUBE NAIL (1″ diameter cap) Tip spreads out when driven (as shown) to provide back-out resistance.							
CAPPED HEAD NAIL (1″ diameter round or square cap) Annular threaded or spiral threaded.							
SPRING STEEL BARBED CLIP Driven through tin disc.							

(Courtesy: National Roofing Contractors Association)

General Guide to Fasteners (continued)

		Light-Weight Concrete Deck	Poured Gypsum Deck	Steel Deck	Wood Deck		Structural Plank Deck	
					Plywood Deck	Wood Plank Deck	Cement-Wood Fiber Deck	Precast Gypsum Concrete Deck
	HOLLOW CONE SHANK FASTENER Electro-zinc galvanized tapered cone shank.							
	SPLIT SHANK FASTENER Electro-zinc galvanized shank spreads out when driven to provide back-out resistance.							
	ROOFING STAPLE Used only for power driven application over tape or through discs.							
	HARDENED SPLIT SHANK NAIL Shank spreads out when driven to provide back-out resistance.		Use tin caps for nail heads smaller than 1″ dia.					Use tin caps for nail heads smaller than 1″ dia.
	HARDENED STEEL SERRATED NAIL (1″ diameter head with grooved shank) Milled and coated to provide back-out resistance.			Use only on first layer of double layer.				
	THREADED SELF-TAPPING SCREW (Shown in seated position) Driven through a minimum 3″ diameter disc.			Use only on first layer of double layer.				

(Courtesy: National Roofing Contractors Association)

Description: This table is primarily for converting customary U.S. units in the left-hand column to SI metric units in the right-hand column. In addition, conversion factors for some commonly encountered Canadian and non-SI metric units are included.

Metric Conversion Factors

If You Know	Multiply By		To Find
Length	Inches ×	25.4[a]	= Millimeters
	Feet ×	0.3048[a]	= Meters
	Yards ×	0.9144[a]	= Meters
	Miles (statute) ×	1.609	= Kilometers
Area	Square inches ×	645.2	= Square millimeters
	Square feet ×	0.0929	= Square meters
	Square yards ×	0.8361	= Square meters
Volume (Capacity)	Cubic inches ×	16,387	= Cubic millimeters
	Cubic feet ×	0.02832	= Cubic meters
	Cubic yards ×	0.7646	= Cubic meters
	Gallons (U.S. liquids)[b] ×	0.003785	= Cubic meters[c]
	Gallons (Canadian liquid)[b] ×	0.004546	= Cubic meters[c]
	Ounces (U.S. liquid)[b] ×	29.57	= Milliliters[c,d]
	Quarts (U.S. liquid)[b] ×	0.9464	= Liters[c,d]
	Gallons (U.S. liquid)[b] ×	3.785	= Liters[c,d]
Force	Kilograms force[d] ×	9.807	= Newtons
	Pounds force ×	4.448	= Newtons
	Pounds force ×	0.4536	= Kilograms force[d]
	Kips ×	4448	= Newtons
	Kips ×	453.6	= Kilograms force[d]
Pressure, Stress, Strength (Force per unit area)	Kilograms force per square centimeter[d] ×	0.09807	= Megapascals
	Pounds force per square inch (psi) ×	0.006895	= Megapascals
	Kips per square inch ×	6.895	= Megapascals
	Pounds force per square inch (psi) ×	0.07031	= Kilograms force per square centimeter[d]
	Pounds force per square foot ×	47.88	= Pascals
	Pounds force per square foot ×	4.882	= Kilograms force per square meter[d]
Bending Moment or Torque	Inch-pounds force ×	0.01152	= Meter-kilograms force[d]
	Inch-pounds force ×	0.1130	= Newton-meters
	Foot-pounds force ×	0.1383	= Meter-kilograms force[d]
	Foot-pounds force ×	1.356	= Newton-meters
	Meter-kilograms force[d] ×	9.807	= Newton-meters
Mass	Ounces (avoirdupois) ×	28.35	= Grams
	Pounds (avoirdupois) ×	0.4536	= Kilograms
	Tons (metric) ×	1000[a]	= Kilograms
	Tons, short (2000 pounds) ×	907.2	= Kilograms
	Tons, short (2000 pounds) ×	0.9072	= Megagrams[e]
Mass per Unit Volume	Pounds mass per cubic foot ×	16.02	= Kilograms per cubic meter
	Pounds mass per cubic yard ×	0.5933	= Kilograms per cubic meter
	Pounds mass per gallon (U.S. liquid)[b] ×	119.8	= Kilograms per cubic meter
	Pounds mass per gallon (Canadian liquid)[b] ×	99.78	= Kilograms per cubic meter
Temperature	Degrees Fahrenheit	(F-32)/1.8	= Degrees Celsius
	Degrees Fahrenheit	(F+459.67)/1.8	= Degrees Kelvin
	Degrees Celsius	C+273.15	= Degrees Kelvin
Application Rate	U.S. gal. per roof sq.[f] ×	0.4075	= Liter per square meter
	U.K. gal. per roof sq.[g] ×	0.4893	= Liter per square meter
Coverage	Sq. ft. per U.S. gallon ×	0.02454	= Sq. meter per liter
	Sq. ft. per U.S. gallon ×	0.02044	= Sq. meter per liter

[a]The factor given is exact
[b]One U.S. gallon = 0.8327 Canadian gallon
[c]1 liter = 1000 milliliters = 1000 cubic centimeters
  1 cubic decimeter = 0.001 cubic meter
[d]Metric but not SI unit
[e]Called "tonne" in England and "metric ton" in other metric countries
[f]0.4075 mm thick
[g]0.4893 mm thick

DIVISION 7 - THERMAL AND MOISTURE PROTECTION

Section Number	Title
07100	**WATERPROOFING**
-110	Sheet Membrane Waterproofing
	Bituminous Sheet Membrane Waterproofing
	Elastomeric Sheet Membrane Waterproofing
	Modified Bituminous Sheet Membrane Waterproofing
	Thermoplastic Sheet Membrane Waterproofing
-120	Fluid Applied Waterproofing
-125	Sheet Metal Waterproofing
-130	Bentonite Waterproofing
-140	Metal Oxide Waterproofing
-145	Cementitious Waterproofing
07150	**DAMPPROOFING**
-160	Bituminous Dampproofing
-175	Cementitious Dampproofing
07180	**WATER REPELLENTS**
07190	**VAPOR RETARDERS**
07195	**AIR BARRIERS**

Broadscope Explanation

07100 — WATERPROOFING

Impervious membranes, coatings, and other materials applied to walls, slabs, decks, and other surfaces subject to continuous and intermittent hydrostatic pressure and water immersion; includes boards and coatings required for protection of waterproofing.

Related Sections:
   Section 07150 - Dampproofing.
   Section 07500 - Membrane Roofing.
   Section 07570 - Traffic Coatings.

Notes: *Protection boards and sheets with integral drainage media are specified in this section; drainage aggregate is specified in Section 02200 - Earthwork.*

07150 — DAMPPROOFING

Materials installed to provide resistance to moisture penetration through foundation walls and similar surfaces subject to high humidity, dampness, and direct water contact, but not subject to hydrostatic pressures.

Related Sections:
   Section 07100 - Waterproofing.
   Section 07180 - Water Repellents.
   Section 07190 - Vapor Retarders.
   Section 07195 - Air Barriers.
   Section 07200 - Insulation: Vapor retarders integral with insulation.

07180 — WATER REPELLENTS

Transparent materials applied to exposed surfaces of masonry, concrete, stucco, and similar materials to provide resistance to moisture penetration.

Related Sections:
   Section 09900 - Painting: Water repellent paints and stains.

07190 — VAPOR RETARDERS

Bituminous, laminated, and plastic vapor retarders applied separately in wall, roof, and floor assemblies to provide resistance to vapor penetration.

Related Sections:
   Section 07195 - Air Barriers.
   Section 07200 - Insulation: Vapor retarders integral with insulation.
   Section 09900 - Painting: Vapor resistant primers.

Notes: *Vapor retarders under concrete slabs on grade are often specified in Section 03300 - Cast-in-Place Concrete. Vapor retarders under roof deck insulation are usually specified in Section 07500 - Membrane Roofing.*

07195 — AIR BARRIERS

Materials installed at walls, roof/wall connections, at perimeter of door and window openings, and at similar locations to provide a continuous impermeable barrier to air infiltration or exfiltration. Air barriers include reinforced rubber and sheet metal membranes, board products with sealed joints, cement parging, and similar products.

Related Sections:
   Section 07190 - Vapor Retarders.

Notes: *Gypsum board wall surfaces, concrete, metal decking, and similar building elements, which also serve as air barriers, are normally specified in those product sections.*

   Products which serve as both air barriers and vapor retarders are normally specified in Section 07190 - Vapor Retarders.

(courtesy of the Construction Specifications Institute's MASTERFORMAT)

DIVISION 7 — THERMAL AND MOISTURE PROTECTION *Continued*

Section Number	Title
07200	**INSULATION**
-210	Building Insulation
	Batt Insulation
	Building Board Insulation
	Foamed-in-Place Insulation
	Loose Fill Insulation
	Sprayed Insulation
-220	Roof and Deck Insulation
	Asphaltic Perlite Concrete Deck
	Roof Board Insulation

Broadscope Explanation

07200 — INSULATION

Insulation including organic and inorganic insulation applied to walls, roofs, decks, perimeter of foundations, and under concrete slabs on grade, including vapor retarders integral with insulation.

Related Sections:
Section 03500 - Cementitious Decks and Toppings: Lightweight insulating concrete.
Section 07190 - Vapor Retarders.
Section 07195 - Air Barriers.
Section 07250 - Fireproofing.
Section 09500 - Acoustical Treatment: Acoustical insulation.
Section 15250 - Mechanical Insulation: Insulation for piping, equipment, and ductwork.

Notes: *Insulation integral to masonry is often specified in Section 04200 - Unit Masonry. Roof deck insulation is usually specified in Section 07500 - Membrane Roofing. Insulating sheathing is usually specified in Section 06100 - Rough Carpentry.*

07240   **EXTERIOR INSULATION AND FINISH SYSTEMS**

07240 — EXTERIOR INSULATION AND FINISH SYSTEMS

Polymer based, elastomeric finish systems and fiber reinforced, polymer modified, cementitious coating and finish systems; applied at site or in shop directly to foam insulation base, netting, lath, and other compatible substrate.

Notes: *Insulation which is a component of this system should be included in this section.*

Thin coat systems are polymer based; hard coat systems are polymer modified.

Section Number	Title
07250	**FIREPROOFING**
-252	Thermal Barriers for Plastics
-255	Cementitious Fireproofing
-260	Intumescent Mastic Fireproofing
-262	Magnesium Oxychloride Fireproofing
-265	Mineral Fiber Fireproofing

07250 — FIREPROOFING

Special coatings, mineral fiber, and cementitious coverings to provide fire resistance to building components.

Related Sections:
Section 06300 - Wood Treatment: Fire retardant treated lumber.
Section 07270 - Firestopping.
Section 09200 - Lath and Plaster: Plaster fireproofing.
Section 09250 - Gypsum Board: Gypsum board fireproofing.
Section 09800 - Special Coatings: Fire resistant paints.

07270   **FIRESTOPPING**

Fibrous Fire Safing
Fire Penetration Sealants
Firestopping Mortars
Firestopping Pillows
Intumescent Firestopping Foams
Silicone Firestopping Foams
Mechanical Firestopping Devices for Plastic Pipe

07270 — FIRESTOPPING

Material installed in cavities, around pipe penetrations, and in other openings in floors, walls, partitions, and other building components to prevent spread of fire and smoke.

Related Sections:
Section 07200 - Insulation.
Section 07250 - Fireproofing.
Section 07900 - Joint Sealers.

Notes: *Barriers of gypsum board, plaster and similar construction for prevention of the spread of smoke and fire are normally specified in those material sections.*

Firestopping and fire safing for mechanical and electrical penetrations of fire rated assemblies are usually specified in Divisions 15 and 16.

Section Number	Title
07300	**SHINGLES AND ROOFING TILES**
-310	Shingles
	Asphalt Shingles
	Fiberglass Shingles
	Metal Shingles
	Mineral Fiber Cement Shingles
	Porcelain Enamel Shingles
	Slate Shingles
	Wood Shingles
	Wood Shakes
-320	Roofing Tiles
	Clay Roofing Tiles
	Concrete Roofing Tiles
	Metal Roofing Tiles
	Mineral Fiber Cement Roofing Tiles
	Plastic Roofing Tiles

07300 — SHINGLES AND ROOFING TILES

Lapped roofing shingles, shakes and roofing tiles, including underlayment and fastening products and methods.

Related Sections:
Section 07600 - Flashing and Sheet Metal.
Section 07700 - Roof Specialties and Accessories.

(courtesy of the Construction Specifications Institute's MASTERFORMAT)

DIVISION 7 — THERMAL AND MOISTURE PROTECTION *Continued*

Section Number	Title
07400	**MANUFACTURED ROOFING AND SIDING**
-410	Manufactured Roof and Wall Panels
	Manufactured Roof Panels
	Manufactured Wall Panels
-420	Composite Panels
-440	Faced Panels
	Aggregate Coated Panels
	Porcelain Enameled Faced Panels
	Tile Faced Panels
-450	Glass Fiber Reinforced Cementitious Panels
-460	Siding
	Aluminum Siding
	Composition Siding
	Hardboard Siding
	Mineral Fiber Cement Siding
	Plastic Siding
	Plywood Siding
	Steel Siding
	Wood Siding
07480	**EXTERIOR WALL ASSEMBLIES**
07500	**MEMBRANE ROOFING**
-510	Built-Up Bituminous Roofing
	Built-Up Asphalt Roofing
	Built-Up Coal Tar Roofing
-515	Cold Applied Bituminous Roofing
	Cold Applied Mastic Roof Membrane
	Glass Fiber Reinforced Asphalt Emulsion
-520	Prepared Roll Roofing
-525	Modified Bituminous Sheet Roofing
-530	Single Ply Membrane Roofing
-540	Fluid Applied Roofing
-545	Coated Foamed Roofing
-550	Protected Membrane Roofing
-560	Roof Maintenance and Repairs
	Roof Moisture Survey
	Roofing Rest urants
07570	**TRAFFIC COATINGS**
-572	Pedestrian Traffic Coatings
-576	Vehicular Traffic Coatings

Broadscope Explanation

07400 — MANUFACTURED ROOFING AND SIDING

Manufactured components of metal, wood, plywood, plastic, mineral fiber-cement, and composite materials forming wall, roof, and fascia surfaces. Includes both structural and insulated panels.

> *Related Sections:*
> *Section 03400 - Precast Concrete: Precast concrete panels.*
> *Section 04200 - Unit Masonry: Preassembled masonry panels.*
> *Section 07600 - Flashing and Sheet Metal: Sheet metal roofing.*
> *Section 08900 - Glazed Curtain Walls: Wall panels integral with curtain wall system.*
> *Section 10200 - Louvers and Vents.*
>
> *Notes:* *Wood, plywood, composition, and plastic sidings are often specified in Section 06200 - Finish Carpentry.*
>
> *Manufactured roof and wall panels which are part of a pre-engineered building may be specified in Section 13120 - Pre-Engineered Structures.*

07480 — EXTERIOR WALL ASSEMBLIES

Systems of conventional components assembled according to standard details to form exterior infill panels and continuous cladding over several stories. Assemblies typically consist of framing, insulation, substrates, and finish surfaces.

> *Related Sections:*
> *Section 06100 - Rough Carpentry: Wood framing and sheathing.*
> *Section 07200 - Insulation: Building wall insulation.*
> *Section 07400 - Manufactured Roofing and Siding: Mineral fiber cement board.*
> *Section 09100 - Metal Support Systems: Metal studs.*
> *Section 09250 - Gypsum Board: Gypsum panels and sheathing.*
>
> *Notes:* *Typically components for exterior wall assemblies are specified in other sections and installation requirements are specified in this section. As an option, components and installation may be specified entirely in this section.*

07500 — MEMBRANE ROOFING

Roofing systems such as built-up bituminous, modified bituminous, elastomeric single-ply, thermoplastic single-ply, roll, fluid applied, and foamed membranes including surfacing materials and coatings. Roof maintenance, composition and elastomeric flashing, walk boards, and other items integral with the roofing membrane are included.

> *Related Sections:*
> *Section 07100 - Waterproofing.*
> *Section 07190 - Vapor Retarders.*
> *Section 07195 - Air Barriers.*
> *Section 07200 - Insulation: Roof deck insulation.*
> *Section 07600 - Flashing and Sheet Metal.*
> *Section 07700 - Roof Specialties and Accessories.*
>
> *Notes:* *Roof deck insulation, vapor retarders, and air barriers are usually specified in this section.*

07570 — TRAFFIC COATINGS

Surface applied waterproofing and elastomeric and composition type membranes exposed to weather and suitable for light pedestrian and vehicular traffic, but not intended for heavy industrial use.

> *Related Sections:*
> *Section 07100 - Waterproofing.*
> *Section 09700 - Special Flooring.*
> *Section 09780 - Floor Treatment.*

(courtesy of the Construction Specifications Institute's MASTERFORMAT)

DIVISION 7 — THERMAL AND MOISTURE PROTECTION *Continued*

Section Number	Title	Broadscope Explanation

07600 **FLASHING AND SHEET METAL**

-610 Sheet Metal Roofing
-620 Sheet Metal Flashing and Trim
-630 Sheet Metal Roofing Specialties
-650 Flexible Flashing
    Laminated Sheet Flashing
    Plastic Sheet Flashing
    Rubber Sheet Flashing

07600 — FLASHING AND SHEET METAL

Shop and field formed sheet metal roofing with waterproof joints such as flat and standing seams and battens; accessories and trim such as gutters, downspouts, scuppers, gravel stops, copings, expansion joint covers, pitch pans, and diverters; other related protective and ornamental sheet metal items; and metal and flexible flashings for roof and wall construction.

    Related Sections:
        Section 07100 - Waterproofing: Sheet metal waterproofing.
        Section 07400 - Manufactured Roofing and Siding: Manufactured roofing panels.
        Section 07700 - Roof Specialties and Accessories: Manufactured roof expansion joint covers.
        Section 15400 - Plumbing: Plumbing drain flashing.

    Notes: *Flexible flashing integral with masonry is often specified in Section 04200 - Unit Masonry.*

        Sheet metal roofing of standard manufactured components is specified in Section 07400 - Manufactured Roofing and Siding.

07700 **ROOF SPECIALTIES AND ACCESSORIES**

-710 Manufactured Roof Specialties
    Copings
    Counterflashing Systems
    Gravel Stops and Fascias
    Relief Vents
    Reglets
    Roof Expansion Assemblies
-720 Roof Accessories
    Manufactured Curbs
    Roof Hatches
    Gravity Ventilators
    Penthouse Ventilators
    Ridge Vents
    Smoke Vents

07700 — ROOF SPECIALTIES AND ACCESSORIES

Roof specialties and accessories of standard manufactured components, both formed and extruded. Includes accessories installed on and in roofing other than mechanical and structural items.

    Related Sections:
        Section 07600 - Flashing and Sheet Metal: Sheet metal expansion joint covers, accessories, and trim.
        Section 07800 - Skylights.
        Section 08650 - Special Windows: Roof windows.
        Section 10340 - Manufactured Exterior Specialties: Steeples, spires, cupolas, and weathervanes.
        Section 15750 - Heat Transfer: Curbs provided with rooftop mechanical equipment.
        Section 15880 - Air Distribution: Curbs provided with mechanical equipment.

    Notes: *Curbs for mechanical equipment are often specified with the equipment in Division 15.*

07800 **SKYLIGHTS**

-810 Plastic Unit Skylights
    Domed Plastic Unit Skylights
    Pyramid Plastic Unit Skylights
    Vaulted Plastic Unit Skylights
-820 Metal Framed Skylights
    Domed Metal Framed Skylights
    Motorized Metal Framed Skylights
    Ridge Metal Framed Skylights
    Sloped Metal Framed Skylights
    Vaulted Metal Framed Skylights

07800 — SKYLIGHTS

Plastic skylights and metal framed skylight assemblies with plastic and glass glazing.

    Related Sections:
        Section 08650 - Special Windows: Roof windows.
        Section 08900 - Glazed Curtain Walls: Sloped glazing and translucent wall and skylight systems.
        Section 13120 - Pre-Engineered Structures: Greenhouses and other glazed structures.

    Notes: *Glazing for skylights may be specified in Section 08800 - Glazing.*

07900 **JOINT SEALERS**

-910 Joint Fillers and Gaskets
    Compression Seals

07900 — JOINT SEALERS

Elastomeric and nonelastomeric sealants, calking compounds, compression seals, joint fillers, and related accessories.

    Related Sections:
        Section 02500 - Paving and Surfacing: Paving joint sealants.
        Section 05800 - Expansion Control: Expansion joint cover assemblies.
        Section 07270 - Firestopping.
        Section 08800 - Glazing: Glazing sealants.

    Notes: *Acoustical sealants are usually specified in Division 9.*

        Firestopping and fire safing for mechanical and electrical penetrations of fire rated assemblies are usually specified in Divisions 15 and 16.

(courtesy of the Construction Specifications Institute's MASTERFORMAT)

Abrasion Resistance
The capacity of a surface to resist the erosive effects of foot traffic and other abrasion.

Adhesion
The binding together of membrane and substrate.

Aggregate
Granular material such as sand, crushed gravel or stone, slag, and cinders used in the manufacturing of concrete, mortar, grout, asphaltic concrete, and roofing shingles. Aggregate is classified by size and gradation.

Alligatoring
A pattern of rough cracking on a coated surface, similar in appearance to alligator skin. Alligatoring of a surface is usually caused by the application of a coating before the previous coat is dry, or by exposing the surface to extreme heat.

Alloy
A homogeneous mixture of two or more metals, the result of which offers certain desirable properties. Alloys are often used in place of pure materials to reduce costs.

Ambient Temperature
The temperature of the environment surrounding an object.

American Society for Testing & Materials (ASTM)
A Philadelphia-based organization established for the testing and development of standards for construction materials.

APP or Atactic PolyPropylene
A bitumen modifying polymer used in built-up roofing.

Application Rate
A unit of measure used to express a quantity of material applied to an area.

Area Divider
An element used in roofing systems that lack expansion joints. Area dividers are composed of double wood members attached to a flashed wood base plate secured to the roof deck. They are used to relieve thermal stress.

Asbestos
A flexible, noncombustible, inorganic silicate-based fiber formerly used in construction for fireproofing and insulation.

Asphalt
A dark brown to black bitumen pitch that melts readily. It appears naturally in asphalt beds, but is also created as a by-product of petroleum processing.

Asphalt, Air-Blown
Asphalt that has had air blown through it at high temperatures, giving it workability for roofing, pipe coating, foundation waterproofing, and other purposes.

Asphalt Felt
Felt impregnated with asphalt and used in roofing and sheathing systems.

Asphalt Mastic
A viscous asphaltic material used as an adhesive, a waterproofing material, and a joint sealant. Asphalt mastic may be poured when heated, but must be mechanically manipulated for application when cool.

Atactic
A natural or synthetic compound with a high molecular weight.

Backnailing
Nailing the layers or plies of a built-up roof to the substrate to help prevent slippage. Backnailing is performed in addition to hot mopping.

Base Flashing
Flashing that covers the edges of a membrane.

Base Ply
Refers to the first ply when it is not part of a shingled system.

Base Sheet
The saturated and/or coated felt sheeting laid on the first ply in a built-up roof system.

Batten
A strip of wood, steel, or aluminum placed over boards or roof structural members to provide a base for the application of wood or slate shingles or clay tiles; a strip applied to prevent wind uplift of single ply membranes; a raised seam in a metal roof.

Bitumen
Any of several mixtures of naturally occurring or synthetically rendered hydrocarbons and other substances obtained from coal or petroleum through distillation. Bitumen is incorporated into asphalt and coal tar and used in road surfacing, roofing, and waterproofing operations.

Bituminous
A term used to describe products such as asphalt and tar that are composed of, similar to, derived from, relating to, or containing bitumen.

Bituminous Emulsion
(1) The suspension of tiny globules of bituminous substance in a water-based solution. (2) An emulsion, the composition of which is the reverse of the above, i.e., a suspension of tiny globules of water in a liquid bituminous substance. This type of bituminous emulsion is applied to surfaces to provide a weatherproof coating.

Bituminous Grout
A mixture of bituminous material and fine sand or other aggregate which, when heated, becomes liquid and flows into place without mechanical assistance. Bituminous grout "air cures" after being poured into cracks or joints as a filler and/or sealer.

Blind Nailing
Nailing performed so that the nailhead cannot be seen on the face of the work.

Blind Rivet
A method of riveting in which a rivet is inserted from one side, fastened to a stem, pulled against the blind side, and snapped off when the rivet is completely formed.

Blister
An undesirable moisture or air bubble, which is often an indication of delamination. Blisters may occur between roofing membranes or between the membrane and substrate.

Bond
The adhesion holding two components together.

Brooming
Pressing and smoothing a layer of roofing material against freshly applied bitumen, usually with a broom, to create a tight and even bond.

BTU
British Thermal Unit; the energy required to raise the temperature of one pound of water by one degree Fahrenheit.

Built-up Roof Membrane (BUR)
A continuous roof covering consisting of sheets of saturated or coated felt, cemented together with asphalt. The felt sheets are topped with a cap sheet or a flood coat of asphalt, which may have a surfacing of applied gravel or slag.

Butyl Rubber
An elastomer with low gas and water vapor permeability used in low-temperature applications (such as cold storage equipment) or water immersion (water storage).

Calender
A machine used to produce sheet, film, and other materials that are typically wound into rolls by passing through three large counter-rotating steel rollers.

Cant Strip
A three-sided piece of wood, one angle of which is square, used under the roofing on a flat roof where the horizontal surface abuts a vertical wall or parapet. The sloped transition facilitates roofing and waterproofing.

Cap Sheet
The top ply of a mineral-coated felt sheet used on a built-up roof.

Capillarity
The force that results from greater adhesion of a liquid to a solid surface than internal cohesion of the liquid itself.

Caulking
The soft, putty-like material used to fill joints, seams, and cracks.

Chlorinated Polyethylene (CPE)
A thermoplastic "uncured" elastomer used as a single-ply roofing material.

Clorosulfonated Polyethylene (CSPE)
A single-ply roofing material which is neither thermosetting nor thermoplastic.

Closure Strip
Material used to fill in gaps between metal panel ribs.

Coal Tar Bitumen
A brown or black bituminous material made from the distillation of coal, and used as a waterproofing material on level or low-slope built-up roofs and around elements that protrude through a roof. Coal tar bitumen has a lower front-end volatility than coal tar pitch.

Coal Tar Pitch
A brown or black bituminous material produced from the distillation of coal. Coal tar pitch is used as a waterproofing material on level or low-slope built-up roofing and around elements that protrude through a roof. Coal tar pitch softens at 129 to 144 degrees Fahrenheit.

Coated Base Sheet (or Felt)
The underlying sheet of asphalt-impregnated felt used in built-up roofing.

Cold-Process Roofing
Built-up roofing consisting of layers of asphalt-impregnated felts that are bonded and sealed with a cold application of asphalt-based roofing cement.

Collector Box
A device located between the gutter and downspout to facilitate water runoff.

Condensation
The conversion of gas to liquid caused by a decrease in temperature or increase in atmospheric pressure.

Coping
The protective top member of any vertical construction, such as a wall or chimney. A coping may be masonry, metal, or wood, and is usually sloping or bevelled to shed water in such a way that it does not run down the vertical face of the wall. Copings often project out from a wall with a drip groove on the underside.

Corrugation
Molding a flat surface into a parallel wave pattern to impart stiffness and provide strength.

Counterflashing
A thin strip of metal frequently inserted into masonry construction and bent down over other flashings to prevent water from running down the masonry and behind the upturned edge of the base flashing.

Course
A layer of any type of building material, such as siding or shingles, applied for purposes of waterproofing.

Coverage
The amount of surface area that may be covered by a unit of building material.

Crack
A fracture or break caused by thermal stress or substrate movement.

Creep
The slow but continuous permanent deformation of a material under sustained stress. In roofing, creep is caused by movement of the roof membrane under stress.

Cricket
A structure superimposed on a roof to facilitate drainage.

Curb
A frame that protrudes above the surface of the roof, at the edge and/or penetrations to facilitate flashing.

Cure
A change in the physical and/or chemical properties of an adhesive or sealant when mixed with a catalyst or subjected to heat or pressure.

Curing Agent
A substance that, when added to a coating or adhesive, increases or decreases the speed of the curing process.

Cutback Asphalt
A bituminous roof coating or cement that has been thinned with a solvent so that it may be applied without heat to roofs or other areas that require sealing. Cutback asphalt is also used for dampproofing concrete and masonry.

Cutoff
A seal that is applied in order to isolate sections of the roof system to prevent, for example, water from running toward unfinished work areas and damaging exposed membranes.

Dampproofing
An application of a water-resistant treatment or material to a surface to prevent the passage or absorption of water or moisture.

Dead Level
Absolutely level, with no pitch or slope.

Dead-Level Asphalt
A grade of asphalt with a softening point of 140 degrees Fahrenheit (60 degrees Centigrade) used on a level or nearly level roof.

Dead Load
A load that is non-moving. In roofing, a dead load may be an air conditioning unit or the roof deck itself.

Deck
(1) An uncovered wood platform usually attached to or on the roof of a structure. (2) The structural system to which a roof covering is applied.

Degradation
Deterioration of a surface caused by heat, light, moisture, or other elements.

Dehydration
The removal of water through absorption or adsorption.

Delamination
Separation of plies of material, as may affect roofing and laminated wood beams, usually through failure of the adhesive.

Dew Point
The temperature at which air of a given moisture content becomes saturated with water vapor. The dew point may also refer to the temperature at which the relative humidity of the air is 100 percent.

Drain
A pipe, ditch, or trench designed to carry away water.

Dropback
Lowering the softening point of bitumen by heating it in the absence of air.

Ductility
A material's ability to withstand stress by expanding, but without recovering its shape upon removal of the stress.

Eaves
(1) Those portions of a roof that project beyond the outside walls of a building; the bottom edges of a sloping roof.

Edge Sheets
Felt strips used to start the felt shingling pattern at the edge of a roof. These strips are cut narrower than the standard width of the full felt roll.

Edge Stripping (Edging Strip)
(1) The application of edge sheets used to cover joints. (2) A plain or molded strip of wood or metal used to protect the edges of panels or doors and/or to conceal laminations in materials such as plywood.

Edge Venting
The systematic placement of openings along the perimeter of the roof to relieve water vapor pressure.

Elastomer
A term describing various polymers which, after being temporarily deformed by stress, return to their initial size and shape immediately upon the release of the stress.

Elastomeric
Refers to the rubber-like quality of an elastomer.

Elastoplastic
A trade description that refers to a resilient substance that returns to its approximate original shape if deformed to within a certain limit.

Elongation (Stretch)
The process by which a material lengthens to accommodate stress or movement.

Embedment
The process by which one material is placed into another to become part of the whole. In roofing, felt is pressed or embedded into hot bitumen to become part of the roofing membrane.

Glossary

Emulsion
(1) A mixture of two liquids that are insoluble so that the globules of one are suspended in the other (such as oil globules in water). (2) The mixture of insoluble solid particles in a liquid, as in a mixture of uniformly dispersed bitumen particles in water. The cementing action required in roofing and waterproofing occurs as the water evaporates.

Envelope
To prevent bitumen leakage, an envelope or felt fold is created by wrapping part of a base felt over the felt plies above it.

Equiviscous Temperature (EVT) Range
The optimum temperature for the application of bitumen. The EVT range for asphalt occurs at a viscosity of 125 centistokes, plus or minus 25 degrees Fahrenheit.

Ethylene Interpolymer Alloy (EIP)
A thermoplastic material used for single-ply roofing.

Ethylene Propylene Diene Monomer (EPDM)
A thermosetting membrane used for single-ply roofing.

Expansion Joint
A gap or joint between adjacent parts of a building structure or concrete work which allows for safe and inconsequential relative movement of the parts, as caused by thermal variations or other conditions.

Exposure
(1) That part of a shingle which is not covered by another shingle. (2) The time during which a roofing component is in contact with the outdoor elements.

Extrusion
Forcing a material, by heat or pressure, through a die.

Fabric
A cloth or textile woven of natural or man-made fibers.

Factory Mutual Engineering & Research Corporation (FM)
A Norwood, Massachusetts-based organization that provides ratings of roofing assemblies on their resistance to fire and wind uplift for use by insurance companies in the United States.

Factory Square
A unit of measurement; 108 square feet or 10 square meters of roofing material.

Felt
A fabric composed of matted, compressed fibers, usually manufactured from the cellulose fibers found in wood, paper, or rags, or from asbestos or glass fibers.

Felt Mill Ream
Also called "point weight", felt mill ream refers to the mass, measured in pounds, of 480 square feet of dry felt.

Fine Mineral Surfacing
A type of roof surfacing material that is inorganic and insoluble in water. At least 50 percent of fine mill surfacing should be capable of passing through a No. 35 sieve.

Fire-Resistance
(1) The capacity of a material or assembly to withstand fire, demonstrated by the material successfully confining a fire and/or continuing to perform a given structural function. (2) According to OSHA, the property of a material or assembly that makes it so resistant to fire that, for a specified time and under conditions of a standard heat intensity, it will not fail structurally and will not permit the material on the other side of the fire to become hotter than a specified temperature.

Fishmouth
A semi-conical opening formed at the cut edge of a shingle.

Flame-spread
The rate at which a flame moves across exposed decking, or the extent to which a product contributes to the spread of fire.

Flashing
A thin, impervious sheet of material placed in construction to prevent water penetration or to direct the flow of water. Flashing is used especially at roof hips and valleys, roof penetrations, joints between a roof and a vertical wall, and in masonry walls to direct the flow of water and moisture.

Flashing Cement
A mixture of solvent and bitumen, reinforced with inorganic glass or asbestos fibers, and applied with a trowel.

Flat Asphalt
An asphalt with a softening point of 170 degrees Fahrenheit (77 degrees Centigrade) that is used in roofing.

Flood Coat
The top layer of bitumen poured over a built-up roof, which may contain embedded gravel or slag as a protective layer. It is typically poured to 60 pounds per square for asphalt flood coating, and 75 pounds per square for coal tar pitch flood coating.

Fluid-Applied Elastomer
An elastomeric coating material applied at an ambient temperature in a liquid state and drying to form a continuous membrane.

Flutter Fatigue
The weakening that results from the wind uplift of a single ply.

Gable
The portion of the end of a building that extends from the eaves upward to the peak or ridge of the roof. The shape of the gable is determined by the type of building on which it is used: triangular on a building with a simple ridged roof, or semi-octagonal on a building with a gambrel roof.

Galvanize
The process of coating iron or steel with zinc, either by immersion or electroplating, to prevent corrosion.

Gauge
The numerically designated thickness of sheet metal.

Glass Fiber Felt
Used in bituminous waterproofing and roof membranes and shingles, glass fiber felt is produced when glass fibers are bonded to a sheet of felt using resin.

Glass Fiber Mat
A thin mat that is produced from glass fibers with or without the aid of a binding agent.

Glaze Coat
(1) The smooth top layer of asphalt in built-up roofing. (2) A temporary, protective coat of bitumen applied to built-up roofing while awaiting top-pouring and surfacing.

Grain
A metric unit of weight; 7,000 grains equal 1 pound.

Gravel
Coarse particles of rock. Gravel is retained by a No. 4 sieve.

Gravel Stop
A metal strip or flange around the edge of a built-up roof. The stop prevents loose gravel or other surfacing material from washing or being blown off a roof.

Gutter
A shallow channel of wood, metal, or PVC positioned just below and following along the eaves of a building for the purpose of collecting and diverting water from a roof.

Heat Aging
Subjecting a material to high temperatures over a period of time to determine any adverse reaction.

"Hot Stuff" or "Hot"
Hot bitumen.

Hypalons (Hypalon Roofing)
An elastomeric roof covering available commercially in liquid, sheet, or putty-like (caulking) consistency in several different colors. Hypalon (a registered trademark) roofing is more resistant to thermal movement and weathering than Neoprene.

Impact Resistance
The ability of a material to resist dynamic loading.

Incline
A slope, slant, or gradient, expressed either as a ratio or a percentage.

Inorganic Material
Substances, the origin and composition of which are not animal or vegetable, nor the products of organic life.

Knot
An imperfection in fabric construction causing surface inconsistency.

Lamination
The process by which a product or material is produced by bonding together several layers or sheets using an adhesive, pressure, or nails or bolts.

Lap
In construction, a type of joint in which two building elements are not butted up against each other, but are overlapped, with part of one covering part of the other.

Leader
(1) In a hot-air heating system, a duct that conveys hot air to an outlet. (2) A downspout.

Live Load
A temporary and changing load superimposed on structural components by the use and occupancy of the building, not including the wind load, earthquake load, or dead load.

Mansard Roof
A roof that has a change of slope on each of the four sides, the lower slope being steeper.

Manufacturer's Bond
A guarantee made by a security company that it will back a manufacturer's liability to finance repairs necessary within a set time period (usually 5, 10, 15, or 20 years).

Mastic
A thick, bituminous-based adhesive used for applying floor and wall tiles. Also, a waterproof caulking compound used in roofing that retains some elasticity after setting.

Membrane
The impervious layer or layers of material used for water control in the construction of a flat roof.

Membrane Reinforcement
Fabrics (woven or nonwoven) that are saturated and embedded in coating to provide strength.

Mil
A measurement of thickness. One Mil equals .001 inch.

Mineral Fiber Felt
Felt that is composed of mineral wool fibers.

Mineral Granules
Stone used to cover cap sheets, roofing shingles, and granule-surfaced sheets.

Mineral Stabilizer
An inorganic, water-insoluble material mixed with bituminous materials.

Mineral-Surfaced Roofing
A type of built-up roofing material named for its granule-surfaced top ply sheet.

Mineral-Surfaced Sheet
Felt coated with asphalt and surfaced with mineral granules.

Modified Bitumen
The addition of plastic or rubber binders to asphalt to improve its performance and weatherability.

Modulus of Elasticity
The unit of stress divided by the unit of strain of a material that has been subjected to a strain below its elastic limit.

Mole Run
A ridge in a roof membrane that is not associated with deck joints or insulation.

Monolithic
A single entity constructed without joints or seams.

Mop-and-Flop
A process of applying roofing material in which the sheets are placed upside down, coated with adhesive, and then flipped over into place.

Mopping
The application of hot bitumen to a roof substrate, using either a mop or a machine.

Nailing
(1) Exposed nailing leaves the nails exposed to the weather. (2) Concealed nailing leaves nails hidden from outdoor weather. *See also* Backnailing.

Neoprene
A synthetic rubber with high resistance to petroleum products and sunlight. Neoprene is used for many construction applications, such as roofing and flashing, vibration absorption, and sound absorption.

Non-Oxidizing Material
A material composed of substances that will not break down when exposed to oxygen; compounds that weather well with limited oxidation, or that do not harden after exposure.

Oil Canning
The waving or buckling of formed sheet metal, such as roofing or siding.

Orange-Peeling
A surface flaw on paint or polyurethane foam (PUF) that leaves the finish pocked with tiny holes resembling citrus skins.

Organic
Matter that can be classified as a plant or animal; any material composed of hydrocarbons.

Perlite
A volcanic glass used as an insulating material or as a lightweight aggregate in concrete, mortar, and plaster.

Perm
A unit used to measure the transmission of water vapor. The formula is: P = Grains of water vapor/square foot x hours x pressure difference in inches of mercury. (1 inch of mercury = 0.491 psi)

Permeance
The capacity of a substance to resist water vapor transmission.

Picture Framing
The forming of ridges in a rectangular pattern on roof membranes that cover insulation or deck joints.

Pitch Pocket
A method of sealing joints in which a flanged metal container is placed around roof penetrations and filled with hot bitumen.

Ply
A layer. In built-up roofing, a ply is a layer of felt.

Positive Drainage
A sufficient roof slope to allow complete drainage of a roof within 24 hours of rainfall.

Primer
A liquid bitumen substance applied to a surface to increase adhesion for later applications.

Rake
(1) To slant or incline from the vertical or horizontal. (2) A board or molding that is placed along the sloping edge of a frame gable to cover the edges of the siding.

Re-entrant Corner
The acute angle or inside corner of a surface that produces concentrations of stress in the waterproofing or roofing membrane.

Reglet
A groove in a wall to receive flashing.

Reinforced Membrane
A membrane that is strengthened with felt, fabric, fiber, or matting.

Relative Humidity
The ratio of the quantity of water vapor actually present to the amount present in a saturated atmosphere at the same temperature, expressed as a percentage.

Resilience
The ability of a substance to resume its original size and shape following the application and removal of stress.

Resin
A natural or synthetic solid or semisolid material of indefinite and often high molecular weight, which has a tendency to flow under stress.

Roll Roofing
Smooth-surfaced or mineral-surfaced coated felts.

Roof Assembly
The various elements of a roof system (including the roof deck) that cover, weatherproof, and insulate the top surface of a building.

Roof System
The elements of a roof assembly, excluding the roof deck.

Rubber
(1) A highly resilient natural material manufactured from the sap of rubber trees and other plants. (2) Any of the various synthetically-manufactured materials with properties similar to natural rubber.

Sag
The unwanted flow or running of roofing material after application.

Saturated Felt
Felt impregnated with bitumen that has a low softening point.

SBCC
Southern Building Code Congress.

SBR
Styrene-butadiene rubber; a material used as a modified bitumen for built-up roofing. *See also* Modified Bitumen.

SBS
Styrene-butadiene-styrene block copolymer; a material used as a modified bitumen for built-up roofing. *See also* Modified Bitumen.

Screen
A metallic plate or sheet, a woven wire cloth, or other similar device, with regularly spaced apertures of uniform size, mounted in a suitable frame or holder, for use in separating material according to size.

Scrim
A coarse, meshed material, such as wire, cloth, or fiberglass, that spans and reinforces a joint over which plaster will be applied.

Sealant
A material used to seal joints where some movement is anticipated.

Self-Ignition Temperature
The temperature at which a material's self-heating properties will ignite without the presence of an outside ignition source. Variables such as size and heat loss conditions determine the minimum self-ignition temperature.

Selvage
An edge or edging that differs from the main part of a fabric or granule-surfaced roll roofing.

Selvage Joint
A lapped joint used with mineral-surfaced cap sheets in roofing. A small part of the longitudinal edge of the sheet below contains no mineral surfacing so as to improve the bond between the lapped top sheet surface and the bituminous adhesive.

Shark Fin
A felt side lap that is curled upward.

Shingle
A roof-covering unit made of asphalt, wood, slate, asbestos, cement, or other material, cut into stock sizes and applied on sloping roofs in an overlapping pattern.

Sieve
A wire mesh screen used to separate material according to size.

Slag
The nonmetallic waste, developed simultaneously with iron in a blast furnace, that consists of silicates and aluminosilicates of calcium and other bases, and is used as a surfacing aggregate.

Slippage
The lateral movement of adjacent roofing plies.

Slope
An incline. The slope of a roof is measured in inches per foot.

Softening Point
The temperature at which bitumen softens or melts, used as an index of fluidity.

Solid Mopping
The application of hot bitumen over an entire roof surface, leaving no area uncovered.

Special Steep Asphalt
An asphalt used in roofing that softens at approximately 220 degrees Fahrenheit (104 degrees Centigrade).

Split
A crack, tear, or separation in a built-up membrane resulting from movement or tensile stresses. *See* Crack.

Spot Mopping
The application of hot bitumen in a circular pattern, leaving a grid of uncovered area.

Sprinkle Mopping
The application of hot bitumen beads strewn in a random manner over a roof substrate.

Square
A quantity of shingles, shakes, or other roofing or siding materials sufficient to cover 100 square feet when applied in a standard manner; the basic sales units of shingles and shakes.

Steep Asphalt
Roofing asphalt having a high softening point, used on surfaces with a steep slope.

Strawberry
A blister in the coating of a gravel-surfaced roof membrane.

Strip Mopping
A method of applying hot bitumen to a roof deck in parallel strips.

Stripping (Strip-Flashing)
The process of taping joints between the adjacent roof and insulation boards using strips of roofing felt.

Substrate
An underlying material that supports or is bonded to another material on its surface. In roofing, the substrate is the material on which the roofing membrane is laid.

Tapered Edge Strip
A strip of insulation used to: (1) make the transition from one layer of insulation to the next, (2) raise the roof around the perimeter and at curbs.

Tar
A dark, glutinous oil distilled from coal, peat, shale, and resinous woods, used as surface binder in road construction and as a coating in roof installation.

Tensile Strength
The maximum unit stress that a material is capable of resisting under axial tensile loading.

Test Cut
A 4" x 40" sample cut from the roof membrane in order to determine the weight of the interply bitumen poundages and to determine the condition of the roof membrane.

Thermal Barrier
An element of low thermal conductivity placed between two conductive materials to limit heat flow.

Thermal Conductance (C)
The rate at which heat flows through a material surface. The formula for thermal conductance is: C = Thermal Conductivity (k) Thickness (in inches).

Thermal Conductivity (k)
The heat energy which will be transmitted by conductance through one square foot of one-inch thick material in one hour when there is one degree Fahrenheit difference across the material. The formula for thermal conductivity (k) is: k = BTU/Square Foot/Inch/Hour/Degrees Fahrenheit.

Thermal Insulation
A material that provides a high resistance to heat flow.

Thermal Resistance (R)
A unit used to measure a material's resistance to heat transfer. The formula for thermal resistance is: R = Thickness (in inches)/k.

Thermal Shock
The subjection of a material or body, such as partially-hardened concrete, to a rapid change in temperature, which may be expected to have a potentially damaging effect.

Thermal (Hot Air) Welding
Adjoining two surfaces through the application of heat provided from an electric heat gun.

Thermoplastic
The property of a substance that allows it to become soft when heated and hard when cooled.

Thermosetting
The property of a substance that allows it to become rigid and nonremeltable as the result of a chemical reaction.

THF
Tetrahydrofuran, a solvent used in welding thermoplastic materials.

Underwriters' Laboratories (UL)
A private, nonprofit organization that tests, inspects, classifies, and rates devices and components to ensure that manufacturers comply with various UL standards. In roofing, the UL is primarily concerned with fire and wind-uplift resistance of roofing materials.

Vapor-Pressure Gradient
A graphic representation of the various water vapor pressures at different points across a roof.

Vapor Retarder
A material that discourages or resists the transfer of water vapor through a roof. A perm rating of 0.5 or less is considered appropriate for vapor retardation in a roofing system.

Vent
An opening through which smoke, ash, vapor, and airborne impurities may be discharged from an enclosed space to the outside atmosphere.

Vermiculite
A group name for certain clayey minerals, hydrous silicates or aluminum, magnesium, and iron that have been expanded by heat. Vermiculite is used for lightweight aggregate in concrete and as a loose fill for thermal insulating applications.

Wind Uplift
The upward force produced as wind blows around or across a structure or an object.

INDEX

Types of roofing, 4-8

U
Underwriters' Laboratory (UL), 4, 13, 162-163
Unit price estimates, 180

V
Vapor barriers. *See* Vapor retarders
Vapor drives, 32
Vapor retarders, 37-43, 42
Ventilation requirements
  for wood decks, 32
Venting, 149
Voids, 65

W
Warranties, 155-158
Warranty
  problems, 157

reliability, 157
Watershed materials, 8
Weather, 18
Wet decks, 65
Wind, 19
Windstorm-rated, 162
Wood decks, 32-34
Wood fiber board, 42
Wood shakes
  three types of, 128
Wood shakes and shingles
  decking, 128-129
  fire resistance of, 128
  flashings, 130
  installation considerations, 129
  sequence, 130
Wood shingles
  physical properties, 128